Multimodality in Language and Speech Systems

Text, Speech and Language Technology

VOLUME 19

The titles published in this series are listed at the end of this volume.

Multimodality in Language and Speech Systems

Edited by

Björn Granström

Department of Speech, Music and Hearing,
KTH Royal Institute of Technology, Stockholm, Sweden

David House

Department of Speech, Music and Hearing,
KTH Royal Institute of Technology, Stockholm, Sweden

and

Inger Karlsson

Department of Speech, Music and Hearing,
KTH Royal Institute of Technology, Stockholm, Sweden

KLUWER ACADEMIC PUBLISHERS
DORDRECHT / BOSTON / LONDON

A C.I.P. Catalogue record for this book is available from the Library of Congress.

ISBN 978-90-481-6024-2

Published by Kluwer Academic Publishers,
P.O. Box 17, 3300 AA Dordrecht, The Netherlands.

Sold and distributed in North, Central and South America
by Kluwer Academic Publishers,
101 Philip Drive, Norwell, MA 02061, U.S.A.

In all other countries, sold and distributed
by Kluwer Academic Publishers,
P.O. Box 322, 3300 AH Dordrecht, The Netherlands.

Printed on acid-free paper

Contents

Contents

PREFACE

This book is based on contributions to the Seventh European Summer School on Language and Speech Communication that was held at KTH in Stockholm, Sweden, in July of 1999 under the auspices of the European Language and Speech Network (ELSNET). The topic of the summer school was "Multimodality in Language and Speech Systems" (MiLaSS).

The issue of multimodality in interpersonal, face-to-face communication has been an important research topic for a number of years. With the increasing sophistication of computer-based interactive systems using language and speech, the topic of multimodal interaction has received renewed interest both in terms of human-human interaction and human-machine interaction. Nine lecturers contributed to the summer school with courses on specialized topics ranging from the technology and science of creating talking faces to human-human communication, which is mediated by computer for the handicapped. Eight of the nine lecturers are represented in this book. The summer school attracted more than 60 participants from Europe, Asia and North America representing not only graduate students but also senior researchers from both academia and industry.

A substantial number of people were involved in the preparation of this book and in the summer school on which it is based. In chronological order we would like to thank Steven Krauwer and Mariken Broekhoven at the ELSNET bureau; the Programme Committee of the school: Niels Ole Bernsen, Gerrit Bloothooft, Paul Mc Kevitt, Koenraad de Smedt, Wolfgang Wahlster and Alex Waibel; and the KTH graduate students: Jonas Beskow, Olov Engwall, Joakim Gustafson, Magnus Lundeberg, Tobias Öhman, Anna Sandell and Kåre Sjölander for all their time and effort spent in planning and realizing the summer school. We would also like to thank all the teachers and the students at the summer school for their interest and enthusiasm, which created such a productive environment. We gratefully acknowledge the financial support from our sponsors, ELSNET, TMR and KTH, and from our supporting organisations, ESCA (now ISCA) and EACL.

A special note of thanks goes to the authors of this book and to the external reviewers for their constructive comments. Finally, we would like to devote a large measure of thanks to Cathrin Dunger who was not only instrumental in the practical organization of the summer school but has also carefully and patiently transformed the different contributions into unified and consistent camera-ready copy.

Björn Granström
David House
Inger Karlsson

Contributors

Jens Allwood	Göteborg University, Sweden
Rashid Ansari	The University of Illinois at Chicago, USA
Niels Ole Bernsen	University of Southern Denmark, Denmark
Tom Brøndsted	Aalborg University, Denmark
Robert Bryll	Wright State University, USA
Susan Duncan	Wright State University, USA
A.D.N. Edwards	University of York, UK
Björn Granström	KTH, Sweden
David House	KTH, Sweden
Inger Karlsson	KTH, Sweden
Lars Bo Larsen	Aalborg University, Denmark
Xin-Feng Ma	Wright State University, USA
Karl-Erik McCullough	University of Chicago, USA
Paul Mc Kevitt	University of Ulster (Magee), Northern Ireland
David McNeill	University of Chicago,USA
Michael Manthey	Aalborg University, Denmark
Dominic W. Massaro	University of California, Santa Cruz, USA
Thomas B. Moeslund	Aalborg University, Denmark
Kristian G. Olesen	Aalborg University, Denmark
Kristinn R. Thórisson	Communicative Machines Inc. New York, USA
Francis Quek	Wright State University, USA

BJÖRN GRANSTRÖM, DAVID HOUSE AND INGER KARLSSON

INTRODUCTION

1. THE ROLE OF MULTIMODALITY IN LANGUAGE AND SPEECH SYSTEMS

What is meant by multimodality and why is it important in language and speech systems? There are a variety of answers to the first part of this question. Definitions of multimodality can be very general, stemming from theoretical models of human information exchange. Other definitions can be based on a particular application framework. In essence, however, multimodality is the use of two or more of the five senses for the exchange of information. From this definition, we can directly infer the importance of multimodality in language and speech systems and provide a number of answers to the second part of the question. First of all, in natural, face-to-face communication, the three sensory modalities of sight, hearing and touch are all used and integrated to a considerable degree. It may be quite important also in human-system interaction to be able to make use of several modalities and modality integration. The use of several modalities may increase the naturalness of human-system communication and in turn lead to more flexible and more efficient systems which are more appealing and easier to use for the general population.

Ease of communication is not the only benefit of using several modalities in language and speech systems. From a system-design point of view, different modes of presentation can complement each other and provide different types of information. They can also reinforce each other and provide more complete information for the user during a difficult task, letting the user make use of different sensory modalities. Here the definition of modality becomes more complex. In his chapter on Modality Theory, for example, Niels Ole Bernsen makes the distinction between media and modality. In the context of this theory, a medium is the physical representation of information. Thus different media are closely related to sensory modalities such as the relationships between graphics and sight, acoustics and hearing, and haptics and touch. Modalities, on the other hand, are defined as different ways or modes of exchanging information. According to this definition, we can talk about different information channels, which can be established between the system and the user. The use of multiple channels is not only of great interest to the general design of systems, but it is also important to many people with sensory impairments as is explained by Alistair Edwards in his chapter on multimodal interaction and persons with disabilities.

Technological advances which have made possible rapid development of new interactive interfaces have also led to a considerable increase in interest in multimedia applications and multimodal information systems in general. The current status of such systems is, however, often more experimental than commercial, and

B. Granström et al. (eds.), Multimodality in Language and Speech Systems, 1–6.

bears witness to the fact that there are still many problems to solve before such complex systems can be truly useful. The question of modality interaction, for example, is crucial to the development of multimodal systems. Which modality is best for a given task and in what situation? How do the different modalities reinforce or complement one another? In what situations or context is it a detriment to use one or the other modality? How do speech and language interact with other modes of presentation, and when is it advisable to use speech? What is the role of animated agents in such systems and how is speech perceived in a multimodal setting? These are just some of the questions that the different authors of this volume discuss and illuminate. The area of multimodality research is broad and comprises many disciplines. This book makes no pretence of covering the area completely, but rather represents selected perspectives on multimodality relevant to language and speech systems and includes several examples of current experimental and research oriented systems.

From the early oral storytelling tradition, where body and facial gestures have always played a major role, through classic theater, opera, dance, film, video and multimedia, multimodal communication has been constantly present in human communication and human culture. As Jens Allwood in his chapter on bodily communication points out, interest in the study of gestures and communication can be traced back to antiquity. Modern studies of multimodal communication, however, have their roots in the work of psychologists and linguistics during the past 50 years made possible by the advent of film and video analysis techniques. This book, therefore, takes as its point of departure, studies of human-to-human communication involving bodily communication (Allwood) and speech and gesture (McNeill et al.). The final chapter of the first section involves audio-visual speech perception (Massaro) where much of what we know is a result of experimentation using visual speech synthesis. Visual speech can also be used by the hearing-impaired to replace or reinforce the impaired modality of hearing. This is a good example of modality transform, which is elaborated on in the section on multimodality in alternative communication mediated by machine (Edwards). That chapter serves as a natural link to the final section on multimodality in communication between humans and systems. The section begins with a taxonomy and systematic analysis of input and output modalities, their representations and functions, and discusses practical problems of modality selection (Bernsen). This modality overview is followed by the final three chapters in which various aspects of experimental language and speech systems are described and tested. The first of these chapters (Brøndsted et al.) presents the architecture of a prototype system 'CHAMELEON' where input includes spoken language and pointing gestures and output comprises speech and a system pointer. The following chapter describes a computational model of natural turn-taking in goal-oriented face-to-face dialogue and presents results based on the implementation and testing of the model (Thórisson). The book concludes with a chapter on visual and auditory conversational signals for artificial talking heads and presents examples from different experimental spoken dialogue systems in several domains (Granström et al.).

2. MULTIMODALITY IN HUMAN-TO-HUMAN COMMUNICATION

2.1. Bodily Communication

A prerequisite to language and speech system design is the understanding of multimodal communication between humans. In his chapter on bodily communication, Jens Allwood presents us with a summary of the history of such research, and then places gestural and bodily communication in a human communication framework where the key concepts of intention and awareness are related to the sharing of different types of information. The chapter continues with examples of how bodily movement can be used as a primary means of expression and a review of some of the functions and content of such movements and gestures in human communication. Finally the chapter concludes with a section on the relationship between speech and gestures exemplifying three types of relationships: addition of information, change of information and reinforcement/support of information; and discusses the interaction between multimodal contributions from different communication partners. This final section is of considerable importance for system designers since one must understand, for example, when and what type of multimodal presentation is beneficial to the transfer of information. The section on the relationship between speech and gestures provides a natural transition to the following chapter on precisely that subject.

2.2. Speech and Gesture

In the chapter by David McNeill, Francis Quek, Karl-Erik McCullough, Susan Duncan, Robert Bryll, Xin-Feng Ma and Rashid Ansari on dynamic imagery in speech and gesture, the guiding argument is that gestures are part of our thinking process and that gestures and speech arise together from the same 'idea units.' They propose that the two modalities are not redundant but rather that each modality expresses its own part of the same semantic idea unit. They also introduce the concept of the 'catchment', which is a term they use for discourse units inferred on the basis of recurring gestural features. Video processing techniques are used to obtain gesture motion traces, which are compared to the fundamental frequency and power amplitude of the audio speech signal. Using the catchment concept, the authors are able to provide experimental evidence showing correlation between discourse segments signalled by gesture and those signalled by speech. The gestural features signalling the discourse segments in the material are explained and it is shown that these features also correspond to an independently derived text-based analysis of the discourse segmentation. The kind of analysis presented in this chapter and the correlation found between speech, gesture and discourse segmentation are important for both multimodal input and output considerations in speech and language systems.

2.3 Audio-visual speech perception

In his chapter on multimodal speech perception, Dominic Massaro presents a number of theoretical arguments and a large amount of empirical evidence supporting an approach to human speech perception as a multimodal phenomenon rather than as primarily an auditory one. One of the main arguments deals with the complementary nature of auditory and visual speech. When information from one modality is weak, it is complemented by stronger cues from the other modality. Massaro places spoken language understanding within an information-processing framework involving the sequential processing of different sources of information. A fuzzy logical model of perception (FLMP) has been developed in which the integration of the auditory and visual information sources takes place in an optimally efficient manner. This model is used to explain a variety of results from several audio-visual perception tests. In the tests, audio-visual synthesis in the form of a talking head is used to create test stimuli in which both audio and visual speech cues are systematically varied. The results of these tests give us insight into how humans combine and integrate audio and visual cues in speech perception. This knowledge is important in designing systems using talking heads as interactive agents. Massaro illustrates this in the context of visible speech in applications for the hearing-impaired. This type of application is the topic of the following section of the book.

3. MULTIMODALTY IN ALTERNATIVE COMMUNICATION

A special situation in multimodal human-human or human-system communication occurs when one or more human modalities are impaired or lost. The impairments can be of different types: sensory impairments such as deafness or blindness, physical impairment such as lack of motor control of the speech or manual articulatory system, or cognitive impairments such as dyslexia. Different ways of solving communication problems that occur in these cases are discussed by Alistair Edwards. Multiple modalities are very important in this context as an impairment in or a loss of one modality may be overcome by the exploitation of another. An example of this is the use of speech output from a computer to assist visually impaired users in reading a computer screen. A partial loss of a modality can also be remedied by enhancing input to that channel.

As is discussed in the chapter, this area of research puts heavy demands on the knowledge and understanding of the inherent differences between the senses and also on the different ways one particular modality can be used. Examples are given of existing and possible communication devices that may be used both in human-to-human and in human-system communication. The aims of these devices are to enhance the quality of life using the concept 'design for all.' The devices are not only useful for people with impairments but may also be useful in particular situations for everybody, for example in noisy environments. The challenges presented by the need for alternative communication are seen to stimulate innovations in system design which will be beneficial to all users.

4. MULTIMODALITY IN HUMAN-SYSTEM COMMUNICATION

4.1. Modality Theory Framework

With the increasing use of different modalities and combinations of modalities in human-system interaction, the need for a taxonomy and a systematic analysis of the modalities has evolved. Niels Ole Bernsen has addressed this challenging task in his chapter and presents a definition and discussion of a Modality Theory. The theory addresses the general problem: given any particular set of information which needs to be exchanged between the user and the system during task performance in a given context, identify the input/output modalities which constitute an optimal solution to the representation and exchange of that information. The taxonomy and theory accordingly provide a framework for describing different modalities and for deciding which modalities to choose or to avoid for a given application. Modality theory is specifically applied to speech functionality where the decision to use speech output and/or speech input in interaction design is discussed. The use of the theoretical framework in an interactive design support tool that assists in choosing between modalities in a specified situation is also demonstrated.

4.2. A Hands-Free System Integrating Language and Vision

Multimodal systems often include screen presentations of data and employ key-boards and screen pointing devices (e.g. mouse) as modes of communication for the user, which means that the user has to stay in front of a screen. This is not always necessary in multimodal systems. An example of an application where the user is not tied to a screen is described in the chapter by Tom Brøndsted, Lars Bo Larsen, Michael Manthey, Paul Mc Kevitt, Thomas Moeslund, and Kristian Olesen. Both input and output to this experimental system are multimodal using speech recogni-tion, gesture recognition, natural language processing, speech synthesis, and a laser pointing system. The user may walk around in a room and ask for information from the system by speaking and/or by pointing at a floor-plan of a building placed on a table. The system response will be by speech and by pointing and tracing with a laser beam either at a specific position or indicating a route between locations on the floor-plan. This chapter gives us an example of the how the integration of language and vision processing in a dialogue system can be achieved and shows us the potential of a hands-free system for information exchange.

4.3. A Computational Turn-Taking Model

Kristinn Thórisson's main interest in this chapter lies in the problem of turn-taking in a multimodal dialogue system, i.e. how should the system signal that it expects reactions from the user and how can the system perceive that the user is expecting the system to react. A good command of this problem will greatly improve the seamless operation of multimodal dialogue systems. Thórisson discusses data from studies of how humans behave in dialogues. He has used this data to develop a computational turn-taking model, which has been implemented using a humanoid agent. The turn-taking cues are multimodal in nature; they contain both speech and non-speech sounds, gestures and body language such as eye-gaze. The duration of these signals and their timing relative to each other and to the dialogue partner's speech and gestures are of crucial importance. The prototype system described in the chapter perceives the user's behaviour and responds with real-time animation and speech output. The system has been tested by hundreds of naive human users. When comparing the system with and without the turn-taking mechanisms it was found that the subjective scores for the system's language understanding and language expression were significantly higher when the turn-taking model was implemented and came close to those given for human-human interaction.

4.4. Speech and Gestures for Talking Head Agents

Björn Granström, David House and Jonas Beskow focus on the use of a 3D interactive interface agent, in most cases a talking head, as a means of facilitating interaction in dialogue systems. They present a method of audio-visual speech synthesis in which synchronized speech gestures and face and head movements are produced by a text-to-speech program. The talking head is used as an output from dialogue systems in combination with other output modalities while the input may be speech, pointing on a screen or typing. The benefits of using a talking head in an interactive system have been investigated and demonstrated in perceptual tests. Benefits include enhanced speech perception and understanding when the talking head is used as speech-reading support for both normal-hearing listeners (speech in noise) and hearing-impaired listeners. Finally, gestures for signalling prominence, emotion, turn-taking, feedback, and system status (e.g. thinking and searching) are explored and discussed in the context of several different applications and domains.

5. CONCLUSIONS

We have so far only seen the beginning of a very interesting development of more human-like and user-friendly multimodal system interfaces. As can be understood from the first chapters of this book, our knowledge of multimodal human-human communication is fairly large and growing. Combining this knowledge with the possibilities generated by technological progress and innovation will lead to more sophisticated and user-friendly system interfaces. This will give us different types of systems that can be used by everybody. The later part of this book attempts to give an idea of how this can be achieved by providing both a theoretical framework and several examples. The examples are all system prototypes that have been used to test different modalities as conveyors of information. The examples far from exhaust the possibilities; rather the intention is to give the reader a taste of what can be done and how this may make communication between humans and systems smoother and easier. Finally, we hope that this book will help inspire new innovations and uses of multimodality in language and speech systems.

JENS ALLWOOD

BODILY COMMUNICATION DIMENSIONS OF EXPRESSION AND CONTENT

1. INTRODUCTION

Bodily communication perceived visually or through the tactile senses has a central place in human communication. It is probably basic both from an ontogenetic and a phylogenetic perspective, being connected with archaic levels in our brains such as the limbic system and the autonomous neural system. It is interesting from a biological, psychological and social point of view and given recent developments in ICT (Information and Communication Technology). It is also becoming more and more interesting from a technological point of view.

However, interest in bodily communication is not new. There is preserved testimony of interest in the communicative function of bodily movements since antiquity, especially in connection with rhetoric and drama (cf. Øyslebø, 1989). However, the study of bodily communication has clearly become more important over the last 40 years, related to an increased interest in the communication conveyed through movies, television, videos, computer games and virtual reality.

In fact, it is only with easily available facilities for recording and analyzing human movements that the study of bodily communication really becomes possible. It is becoming increasingly important in studies of political rhetoric, psychodynamically charged communication and communication in virtual reality environments. Pioneers in the modern study of bodily communication go back to the 1930's when Gregory Bateson filmed Communication on Bali (cf. Lipset, 1980) or the 1950's when Carl Herman Hjortsjö (e.g. Hjortsjö, 1969) started his investigations of the anatomical muscular background of facial muscles, later to be completed by Paul Ekman and associates (Ekman & Friesen, 1969). Another breakthrough was made by Gunnar Johansson (e.g. Johansson, 1973) who, by filming moving people dressed in black with white spots on their arms and legs, was able to make a first attempt at isolating what gestures are significant in communication. Other important steps using filmed data were taken by Michael Argyle (1975), Desmond Morris (1977), Adam Kendon (1981) and David McNeill (1979). In the 1990's, another barrier was crossed when it became possible to study gestures using computer simulations in a virtual reality environment (cf. Cassell et al, 2000).

For an overview of the whole field and its development there are several introductions available. Among them are Knapp (1978 and later editions), Key (1982), Øyslebø (1989) and Cassell et al (2000).

B. Granström et al. (eds.), Multimodality in Language and Speech Systems, 7–26.

2. THE PLACE OF BODILY COMMUNICATION IN HUMAN COMMUNICATION

2.1 Communication

If we try to define the word communication in a way, which covers most (perhaps all) of its uses, we get a definition of the following type:

Communication = def. Transmission of content X from a sender Y to a recipient Z using an expression W and a medium Q in an environment E with a purpose-/function F.

Even if it is possible to add further parameters, some of the most important are given in the above definition. The definition could be paraphrased by saying that communication in the widest sense is *transmission of anything from anything to anything with the help of anything (expression/medium) in any environment with any purpose/function.* A definition which is as wide as this is required to capture uses of the word *communication* which are exemplified in expressions like *table of communication, railroad communication* and *communication of energy from one molecule to another* (cf. Allwood, 1983).

Based on these examples, it could be claimed that the word communicant designates a "pretheoretical concept" which needs to be made more precise and specific in order to be suitable for theoretical analysis. This could, for example, be done by analyzing the connections and relations between properties of the arguments in the definition that provide constraints and enablements, i.e. properties and relations of the content (X), the sender (Y), the recipient (Z), the expression (W), the medium (Q), the environment (E) and the purpose/function (F).

Some of these properties and relations are the following:

1. *Sender and recipient*: A first problem here concerns the terms *sender* and *recipient.* Depending on circumstances, the following terms could be used as synonyms of *sender: speaker, communicator, producer, contributor* and the following as synonyms of *recipient: listener, hearer, communicator, receiver, contributor*. All terms have problems since they are either too restricted, too general (no difference between sending - receiving) or give the wrong metaphorical associations – *sender* and *receiver* are too closely linked to radio signaling. A second problem concerns how the nature of senders and recipients influence their ability to communicate. Some of the most important abilities of senders and recipients have to do with whether they are living, conscious and capable of having intentions. Their abilities often relate to what types of causal and social relations they have to their environment. Different types of senders and recipients vary greatly in their ability to make use of such relations in order to convey and receive information symbolically, iconically and indexically. See section 2.2 below.

2. *Expressions and media:* Which types of expression and media are available to senders and recipients depends on the restrictions and enablements that are imposed by their nature. Through their five senses, human beings can perceive

causal influences of at least four types (optical, acoustic, pressure and chemical (taste, smell). These causal influences have usually been produced by bodily movements or secretions coming from other human beings. Normal human face-to-face communication is, thus, multimodal both from the point of view of perception and production, employing several types of expression and media simultaneously.

3. *Content*: Similarly, the content is usually multidimensional. It is often simultaneously factual, emotional-attitudinal and socially regulating. There are several interesting relations between the modalities of expression and the dimensions of content, e.g. we mostly communicate emotion using vocal quality or body movements while factual information is mostly given with words.

4. *Purpose and function*: On a collective, abstract level, the purposes/functions of communication can, for example, be physical, biological, psychological or social, e.g. "survival" or "social cohesion". On a more concrete level, most individual contributions to conversation can also be connected with (individual) purposes/functions, like making a claim or trying to obtain information.

5. *Environments*: Environment on a collective, abstract level can be characterized as physical, biological, psychological or social in a way which is similar to "purpose/functions". Each type of environment can then be connected with particular types of causal influence in communication. On a concrete level, most human environments will be complex combinations of all the four mentioned dimensions and possibly others and thus exert a fairly complex combined influence on communication.

2.2 *Indices, icons and symbols*

People who communicate are normally situated in a fairly complex (physical, chemical, biological, psychological and social) environment. Through their perception (i.e. at least sight, hearing, touch, smell and taste) connected with central brain processing, they can discriminate objects, properties, relations, processes, states, events and complex combinations of all of these in their environment. All information, including that originating in communication with other persons, is processed and related to preexisting memories, thoughts, emotions or desires and in this way makes up a basis for what later can be expressed in communication.

What a person expresses can normally be described as being dependent on the attitudes the person has toward the expressed information. Clear examples of this can be found in such speech acts as statements, questions and requests, which normally express the cognitive attitudes of belief, inquisitiveness and desire for some action on the part of the hearer.

Independently of what is going to be expressed, any communicator has to use one of three basic ways of conveying and sharing information (cf. C.S. Peirce,

1902). Peirce was concerned with a general basic descriptive framework for communication and sharing of all types of information (including information related to gestures), so his "semiotics" contains many concepts, which are useful in describing multimodal communication:

A. *Indexical information*; this is information which is shared by being causally related to the information which is being perceived - the index, e.g. black clouds, can be an index of rain.

B. *Iconic information*; this is information which is shared by being related through similarity or homomorphism to the information which is being perceived - the icon, e.g. a picture, iconically represents whatever is depicted.

C. *Symbolic information*; this is information which is shared by being related by social convention to the information which is being perceived – the symbol, e.g. words, symbolically represent their referents.

In normal human communication, we simultaneously use a combination of these types of information. For example, as we speak to each other, we frequently let our words "symbolically express" factual information while our hands "iconically illustrate" the same thing and our voice quality and our facial gestures "indexically" convey our attitude to the topic we are speaking about or the person we are speaking to.

The simultaneous and parallel use of symbolic, iconic and indexical information is commonly connected with variation in the extent to which we are aware of what we are doing and variation regarding how intentional our actions are. Generally we are most aware of what we are attempting to convey and share through symbols, somewhat less aware of what we convey and share iconically and least aware of what we convey and share indexically. This means that most people are more aware of what they are trying to say than they are of what their hands illustrate or of what their voice quality and facial gestures express.

This variation in intentionality and awareness also leads to a variation in controllability which affects our impression of how "authentic" or "genuine" the feelings and attitudes of a person are. Usually this impression is more influenced by voice quality and gestures which are not easily controllable than by those that are more readily controllable.

If a conflict arises between what is expressed by words or by facial gestures which are relatively easy to control and what is expressed by voice quality or by the rest of the body, which is not so easy to control, we mostly seem to trust information which is not so easy to consciously control. More or less subconsciously, we seem to assume that such information puts us in touch with more spontaneous, unreflected reactions.

However, this tendency has sometimes been misunderstood in previous research on nonverbal communication (cf. e.g. Fast, 1973). The significance of what has been said above is not that 80-90% of the information that is shared in conversation is conveyed by bodily movements. The significance is not even that information which is conveyed by bodily movements is more important than other types of information.

Rather the significance is that bodily movements and voice quality are convenient, spontaneous and automatic means of expression for emotions and attitudes. Probably, they are our most important means of expression for this type of information. As a consequence they often also become our most genuine and spontaneous means of emotional expression. However, this does not imply that information about emotions and attitudes is always the most important information. Sometimes it is, sometimes it is not - sometimes factual information is more important. Nor does it imply that genuine or spontaneous expression of emotion is always the most appropriate or the most interesting.

An emotional expression based on some effort and reflection can in certain situations be more interesting and appropriate. After all, this is what the person wants to express and leave as a lasting impression, using effort, self-control and reflection.

2.3 *Indicate, display and signal*

Above I have briefly illustrated that one of the interesting questions connected with the study of how body movements are used for communication is the question of how intentional and conscious or aware such communication is. Since this problem is of both theoretical and practical interest, I will now introduce three concepts which can be used to capture some of the variation in degrees of intentionality and awareness (cf. also Allwood, 1976 and 2000, as well as Nivre, 1992).

A. *Indicate*: A sender indicates information to a recipient if and only if he/she conveys the information without consciously intending to do so. If A blushes in trying to answer a sensitive question this could indicate to the recipient that A is feeling shy or embarrassed. Information that is indicated is thus causally connected with A without being the product of conscious intention. It is totally dependent on the recipient´s ability to interpret and explain what A is doing.

B. *Display*: A sender displays information to a recipient if and only if he/she consciously shows the information to the recipient. For example, a person A can consciously use more of his/her regional accent in speaking in order to show (display) where he/she is from.

C. *Signal*: A sender signals information to a recipient if and only if he/she consciously shows the recipient that the information is displayed. To display is to show that you are showing. Ordinary verbal communication usually involves signaling. For example, if a person A says *I am from Austin* this information is signaled, i.e. it is clear that the sender wants the recipient to notice that he/she is communicating (showing) this information.

The three concepts *indicate*, *display* and *signal* are really three approximate positions on a complex scale combining degrees of consciousness and intentionality. "Indicate" is connected with a lack of conscious intentionality while "display" and "signal" are associated with greater degrees of awareness and intentionality. However, consciousness and intentionality are in themselves very complex phenomena so that the three concepts only capture some of their properties. Other concepts might be needed to capture other types of intentional and conscious states than the ones described here, e.g. the higher levels of iterated (reflexive) consciousness and intentionality described in Schiffer (1972). The only claim made here is that the three concepts can be a useful point of departure for a description of consciousness and intentionality in communication.

It is possible to combine the three types of communicative intentionality and awareness with the three basic semiotic relations described earlier (indexical, iconic and symbolic). If we do this, we obtain a table with the following nine combinations:

Table 1. Indices, icons, symbols and degrees of communicative awareness and intentionality

	Index	*Icon*	*Symbol*
Indicate	X		
Display		X	
Signal			X

All combinations are possible in principle, but in practice certain combinations are more common than others. In the table, an X has marked this. For example, indexical information is mostly indicated (this was in fact the motivation for the choice of the term "indicate"), even if with conscious, intentional effort it can be displayed and/or signalled. For example, we might with the help of bio-feedback learn how to blush. Similarly, symbolic information is mostly signalled even if it can also be communicated with a lower degree of consciousness and awareness. Iconic information is mostly displayed but can exceptionally be indicated. The reason for these preferential relations is thus far not fully clarified. It involves, for example, looking at whether iconic (isomorphic) relations are more easily usable in the visual than the auditive mode and whether visual icons are more suitable for display than for signalling.

As we have seen, normal human face-to-face communication is multidimensional. Among other things this means that the source of shared information can be indexical, iconic and symbolic, and that the sharing simultaneously can occur on several levels of intentionality and awareness by being indicated, displayed or signalled. Normal (multimodal and multidimensional) communication thus carries with it the complex task of integrating diverse modalities and levels of awareness into the complex resulting shared content.

In this way, normal human communication just like the communication of other species contains much sharing of information on an indicated and displayed level.

This kind of information forms a common basis for communication across species. What differentiates humans from other species, as far as we know, is the large-scale introduction of signalled symbolic communication and the high degree of complex use of several levels of communicative intentionality simultaneously.

This view should be contrasted with a traditional linguistic perspective which usually assumes that linguistic communication is only signalled and symbolic (mostly in written form). The insufficiency of this perspective becomes apparent as soon as we start to seriously describe spoken language communication and include intonation and bodily movements in the description. We then notice that normal spoken interaction, besides being symbolic (digital), also is iconic and indexical (analog), and that this information can be shared not only through signalling but also by being displayed or indicated.

We will also notice that reception and sharing of information is neither passive, nor always conscious. Reception, i.e. perception and understanding (if we want one word we can use J.L. Austin's word "uptake", (cf. Austin, 1962) is dynamic just like production (sending) of information, being controlled by perspectives and purposes which are often unaware, so that a person reacts and stores information in an automatic way without being fully conscious of what is happening.

Body movements and prosody are thus very important means of displaying & indicating indexical and iconic information simultaneously with signalled symbolic verbal information. It should, however, be noted that the major focus of intentional effort can be changed so that symbolic information can be used with a low degree of intentionality and awareness and indexical information with a higher degree of awareness. It should also be stressed that bodily movement often can be used to convey symbolic information. Deaf sign language very clearly shows this. Let us now, in more detail, consider the different means of expression employed in communication.

3. MEANS OF EXPRESSION IN COMMUNICATION

The means by which humans communicate can be subdivided in many ways. One possibility is the following:

A. *Primary:* Primary means of expression are means of communication that can be controlled directly without extra aids, e.g. bodily movements, voice, speech, gestures, touch, song, etc. Possibly production of molecules related to smell and taste could also be included. An argument against including smell and taste is that even though they are directly causally related to man, they are usually not controllable. Concerning the other primary means of expression, they include both spontaneous indexical and iconic means as well as symbolic means dependent on social conventions (speech, gestural language and certain types of song).

B. *Secondary*: Secondary means of expression simply consist of the instruments which are used to augment and support the primary means of expression.

Secondary means are used, for example, to overcome spatial distance and to preserve information over time, e.g. using pen, chisel, typewriter, computer, megaphone, microphone with a loudspeaker, semaphore, radio, TV, audio and videotapes, telephone, telegraph, fax or e-mail. As we can see some secondary means directly reproduce primary means, e.g. radio, megaphones, audio tapes while others require more advanced recoding of primary means, e.g. writing. In some cases, this recoding requires several steps, e.g. telegraph or e-mail.

C. *Tertiary*: When we come to tertiary means it might be objected that the label "means of expression" is not entirely adequate. Tertiary means are simply all human artifacts (no negative evaluation intended) that are not secondary means of expression, e.g. tables, chairs, houses, roads, household appliances, cars, etc. All such artifacts express technical, functional and aesthetic ideas and intuitions. Perhaps the artifacts which are easiest to regard as means of expression are those which mainly have an aesthetic purpose like paintings and sculptures, etc. Second to these, there are artifacts, the construction and shaping of which has been under relatively direct causal control by the person who has made them. In most traditional cultures involving "handicraft", such control was usually individually exercised by both masters and apprentices. The artifacts in industrial societies, however, have less and less of such individual control and are instead often products of teamwork and industrial mass production. If they are to be seen as means of expression they must perhaps be seen as an expression of a collective rather than an individual mentality. In fact, this was perhaps also true of older traditions and artifacts where the creation of a single individual often was constrained by tradition and for this reason difficult to discover.

Thus, tertiary means of expression can often be regarded as collective while primary and secondary means, even if they are also often bound by convention, give greater room for the expression of single individuals.

What we have here been calling tertiary means of expression could also be extended to include the unintended and undesired remains that different human cultures have left behind, e.g. bits of pottery, charcoaled remains of houses, leftovers from eating and more generally a changed and somewhat destroyed environment. All of them are in extended sense expressions of human activities and tell us something about the collective forms of life that produced them. Since some of them might be intended while others probably are unintended, we see that also with regard to expressions of collective forms of life there are varying levels of awareness and intentionality. Both collective and individual expressions can be indicated, displayed or signalled, and both types can make use of indexical, iconic or symbolic information.

4. BODILY MOVEMENTS

Let us now turn to movements of the body as primary means of expression and study some of their functions in human communication.

We may first note that body movements can be used both together with speech and independently of speech. They are thus a major source of the multimodal and multidimensional nature of face-to-face communication. Below, we will first discuss some of the major types of body movements and their functions and content, and then return to the question of how they are related to speech.

Some of the body movements that are relevant for communication are the following:

(cf. also Argyle, 1975; Knapp 1978, and later editions; Allwood, 1979; and Øyslebø, 1989). Each type of body movement will be followed by a short description of one or more functions that the movement may have.

(i) Facial gestures. Functions: e.g. emotions and attitudes.

neutral happy sad angry diabolical sheepish

(ii) Head movements. Functions: e.g. information about feedback (acknowledging, agreeing and rejecting) and turntaking, i.e. basic functions for managing interactive dialog and communication.

(iii) Direction of eye gaze and mutual gaze. Functions: e.g. information about attitudes like interest and interactive communication management functions like speaker change (cf. Duncan & Fiske, 1977)

(iv) Pupil size. Functions: e.g. increased pupil size can indicate increased interest.

(v) Lip movements. Functions: e.g. speech or attitudes like surprise.

(vi) Movements of arms and hands. Functions: arm and hand gestures are often used for symbols, e.g. "money" (rubbing thumb against index finger) or "come here" (waving fingers towards palm of hand upwards or downwards depending on culture). They are also used for nonconventional iconic illustrations.

(vii) Movements of legs and feet. Functions: e.g. to indicate nervousness or to display or signal emphasis.

(viii) Posture. Functions: Information about attitudes like shyness or aggression.

(ix) Distance. Functions: information about attitudes. A small distance between communicators could for example indicate friendliness and "closeness".

(x) Spatial orientation. Functions: e.g. information about attitudes like avoidance or contact.

(xi) Clothes and adornments: Functions: e.g. to indicate or display social status or role in a particular social activity.

(xii) Touch: Functions: Touch can be a way of communicating friendliness or aggression.

(xiii) Smell. Functions: Smell can indicate emotional states like fear, what kind of work you do or what food you have been eating. It can also be used to arouse pleasure, displaying a wish to be attractive.

(xiv) Taste. Functions: e.g. information guiding a hungry person in choice of food. Probably taste, if at all used as a means of communication, is used in connection with preparation and consumption of food.

(xv) Nonlinguistic sounds. Functions: e.g. warnings, summons or information about specific types of activity or about specific tasks within an activity.

All examples above are given from the perspective of a producer of the information. From the perspective of the responding recipient, we may note that the majority of the body movements are connected with visual reception (i - xi): (xii) is connected with touch and (xv) with hearing. Smell (xiii) and taste (xiv) are in ordinary language more or less neutral with regard to production and perception. *He smells* can mean both "he is experiencing a smell" and "he is giving off a smell". However, since we are often concerned with human experiencers of smell and taste, phrases like *it smells*, and *it tastes* are often used for production while the recipient side can be described by phrases like *experiencing a smell of X* or *experiencing a taste of X*.

Furthermore, it should be stressed that the functions given above are only meant as examples. There is much more to say. It should also be stressed that the cultural variation both with regard to means of expression and type of function is considerable for almost all of the mentioned types of body movements. Cultural variation is especially well studied with regard to facial gestures, head movements, gaze, arm and hand movements, distance, spatial orientation, clothes and adornments as well as touch.

5. DIMENSIONS OF CONTENT

The use of body movements in communication is typically connected with simultaneous multidimensionality, both with regard to means of expression and functional content. This multidimensionality of body and speech is further connected with differences in levels of awareness and intentionality and with differences in the use of semiotic relations (indexical, iconic and symbolic). Below I will now give a brief account of the dimensions of content (functionality) which are primarily associated with body movements.

1. *Identity:* Movements of the body and the body itself indicate, display and signal who a communicating person is biologically (e.g. sex and age), psychologically (e.g. character traits such as introvert or extrovert) or socioculturally (e.g. ethnic/cultural background, social class, education, region or role in an activity).

2. *Physiological states:* Physiological states of a more or less long-term character, like hunger, fatigue, illness, degree of athletic fitness etc. are often clearly expressed by body movements, e.g. by properties like intensity and agility.

3. *Emotions and attitudes*: When we communicate with other people we continuously express our emotions and attitudes to the topic about which we are communicating as well as to the person with whom we are communicating. We do this primarily with body movements but also with intonation and prosody.

4. *Own communication management*: A fourth function for which we use our body movements is that of managing our own communication. When we need time to reflect, plan or concentrate, we can, for example, turn our gaze away. If we have difficulties finding a word, we often move our body, especially the hands to gain time and to contribute to activating the word (cf. Ahlsén, 1985; Ahlsén, 1991; and Fex & Månsson, 1998). If we need to change what we have said, we may show this by movements of the hands and/or head (cf. also Allwood, Nivre & Ahlsén, 1990).

5. *Interactive communication management:* We also use body movements to manage our interaction with fellow communicators, e.g. Hirsch (1989), based on observation of TV-debates, claims that changes in bodily position can function to show that there is no more to say about a particular topic. Body movements (primarily hands, head and gaze) are also important to regulate turntaking (cf. Duncan & Fiske, 1977; and Sacks, Schegloff & Jefferson, 1975). They are used for feedback, i.e. using facial gestures and head movements in order to show whether we want to continue, whether we have perceived and understood and how we react to the message which is being expressed (cf. Allwood, 1987; and Allwood, Nivre & Ahlsén, 1992). A further important function which perhaps also is primarily managed through body movement is the rhythm of the interaction (cf. Davis, 1982).

6. *Factual information*: Also factual information can be conveyed through body movements. In its most salient form it can be done through the use of symbolic gestures, e.g. in deaf sign language. But symbolic gestures are used also in relation to speech. Probably around fifty symbolic gestures are used together with most of the spoken languages of the world. Some examples are different kinds of head movements for "yes" and "no" (several different cultural variants

exist), shoulder shrugs for "I don't know" and rubbing the index finger against the thumb for "money", a great variety of insulting gestures etc.

In addition, factual information is often conveyed by iconic gestures, so-called "illustrators" (cf. Ahlsén, 1985; and Ekman & Friesen, 1969). It is also conveyed through indexical gestures like pointing or by movements which serve to mark structure or emphasis in the message which is being communicated.

To sum up, we have noted that body movements which are used for communication are multidimensional, both from an expression-oriented behavioral perspective and from a content-oriented functional perspective. Perhaps the most important content-related contributions given by body movements in spoken interaction (between hearing, non-deaf communicators) are related to information about emotions, attitudes and management of interaction.

6. MULTIDIMENSIONAL RELATIONS AND INTERACTION

6.1 The relation between expression and content

Both when they are used on their own and in connection with speech, body movements provide a multidimensional medium of expression which can be used to convey a multidimensional content. The relation between expression and content can thus be described as a simultaneous multidimensional coupling. Consider the following example of how a simple verbal *yes* can be used together with head movements and facial gestures.

	Expression	*Content*
verbal:	yes	affirm
head movement:	nod	affirm
facial gestures:	raised eyebrows	surprise
	wrinkled forehead	doubt

The example shows how affirmation, surprise and doubt can be simultaneously conveyed by a combination of words, head movements and facial gestures.

Secondly, the relation between expression and content is generally a many-to-many relation. Several expressive means are often related to one content - e.g. intensive nodding and an emphatically pronounced *yes* both simultaneously express strong affirmation. Correspondingly, one means of expression can be a codeterminant for many types of content. To return to the example above, nodding can simultaneously signal affirmation and enthusiasm. Thus, there are in general no simple relations between expression and content but many-to-many relations.

Thirdly, the multidimensional and many-to-many coupling between expression and content can take place on several levels of awareness and representation simultaneously. A person can signal something by using conventional symbols, while simultaneously displaying something using similarity or indicating something causally. Natural biological expressions and conventional expressions can be combined and be used together in order to convey different types of content on different levels of awareness and intentionality simultaneously.

Fourthly and finally it is important to note that the perception and understanding of (i) means of expression, (ii) dimensions of content and (iii) the relation between expression and content is dependent on context. Exposed teeth and retracted lips will be seen and understood as a smile or as something else, depending on the look of other facial gestures and the eyes. In order to be interpreted, a single expressive feature must be seen in relation to a surrounding context, e.g. the function of a smile may vary with context in expressing neutrality, ingratiation or shyness.

The socio-cultural context is often decisive in choosing between interpretations of the type mentioned above. In a conversation between two young people freshly in love, one might be more tempted to interpret a smile as shyness than if one is observing a conversation between two older people of different social status, where one might instead be tempted to use the socially stereotyped interpretation - ingratiation.

6.2 The role of body movements compared to prosody

I will now somewhat speculatively compare the contribution given by body movements to human direct "face-to-face" communication with the contribution given by prosody (i.e. variations in the pitch, intonation and intensity of speech), words and grammar. The discussion will use the table below as a point of departure. The types of content that occur in the table are the same as those that have been discussed above, except that focusing has been distinguished as a category of its own and that contextual dependency (which is not really a type of content) has been added.

Table 2. Content/functions which can be expressed by body movements, prosody, words and grammar

Content	*Body movement*	*Prosody*	*Words & grammar*
Identity			
Physiological state			
Emotions, attitudes			
Own communication management			
Interactive communication management			
Factual content			
Focusing			
Contextual dependency			

Let us now consider the types of content one by one. There is probably no great difference between the three means of expression with regard to the possibilities of expressing social identity. This can be done implicitly and indexically using body movements and prosody, or more explicitly and symbolically using words and grammar. Compare the difference between indicating or displaying membership in the upper class implicitly by gestures and prosody and explicitly saying "I am a member of the upper class".

The next category - "physiological states" - is primarily indexically expressed through body appearance and body movements. The third category - emotions and attitudes - is probably mostly communicated through body movements and prosody even if it is clear that words and grammar can also be used to convey emotions and attitudes, especially in poetry.

Turning to "own communication management", (e.g. the ways in which we communicate needs for planning, choice of words and hesitation or the ways in which we show that we want to change what we have said), we probably use all three types of expressive means equally much. This is also true of the ways in which we convey "interactive communication management", i.e. turn management, feedback, sequencing, etc. All three means of expression are used simultaneously, providing information related to more than one type of content.

As for factual content, it seems clear that words and grammar are the most important means. In deaf sign language, bodily gestures replace spoken words and grammar but in ordinary spoken language communication, only a relatively limited number of gestures with a factual content occur. Prosody can play a role for factual information, for example, by being used to make conventionalized distinction between meaningful units like morphemes or words, e.g. by word tones or word accents in many of the languages of the world.

If we consider the structuring of information through focusing, all three types of expressive means may be used. Compare *It was not Bill that Mary kissed*, where *Bill* has been focused grammatically/syntactically with *Mary kissed* <u>*Bill*</u> (giving Bill extra stress), where *Bill* has been focused prosodically. Even though body movements can also be used to emphasize and focus, they probably are less important than spoken words, grammar and prosody.

Finally, the table reminds us that all types of expressive means are dependent on context both in order to be identified as specific types of expression and in order to help us identify what content they are expressing.

6.3 The semiotic status of the production modalities of communication

Maintaining the somewhat simplified 3-part division of the production modalities of communication into "body movement", "prosody" and "words and grammar", we may ask how it relates to the three basic sign types (index, icon and symbol) and to the three types of communicative intentionality (indicate, display and signal) introduced above. Using the preferential relations between indicate and index, display and icon and signal and symbol as shown in Table 1, we may create the following table.

Table 3. Modality of production and semiotic status

	indicate (index)	display (icon)	signal (symbol)
Body movements	X	X	
Prosody			
Words and grammar			X

As in the discussion of Table 1, we may observe that although all combinations are possible, certain relations are preferred among hearing people in "face-to-face" communication.

Words and grammar normally have the status of signalled symbols while body movements mostly indicate (as indices) or display (as icons) information. This is of course very different in deaf sign language where gestures are the main mode of signalled symbolic communication. The status or prosody is more unclear. It clearly very often functions to indicate (as an index) information but it also has important displayed iconic and signalled symbolic functions.

6.4 The relation between speech and gestures

Besides considering the general relation between content and means of expression and the more particular relation of prosody to body movements, it is also of interest to consider the relation between information conveyed by speech and information conveyed by gestures more generally.

Since the two means of expression are separately controllable, the messages they convey can either be independent or dependent. If they are independent, each means of expression carries its own message, e.g. when speaking on the phone and gesturing something to a person in the room simultaneously. If they are dependent, the two means of expression multimodally combine to form a more complex message drawing on both.

If they are dependent, very often, but not always, the spoken message is the main message which the gestural message modifies, e.g. to reinforce some part of what has been said. Sometimes, however, the gestural message might be more important, as when a person exhibits a particular emotion through posture and facial expression and words only serve to fine-tune the emotion. There are also cases when the two are more or less of equal importance, e.g. (in giving directions on how to find something) saying *over there* and accompanying this utterance with a pointing gesture showing the exact location.

More generally, when speech and gesture are not used for messages which are independent of each other, the relation between them can be of three kinds:

(1) Addition of information:
- identity expression (anchoring)
- attitudinal embedding
- illustration
- specification
- communication management

(2) Change of information:
- Attitudinal modification
- Communication management

(3) Reinforcement/support of information:
- Support, repeat

Let us now consider the three kinds of relations a little more carefully one by one:

1. *Addition of information*: The first type of relation involves one means of expression adding information to the information given by the other means – for example, when a speaker, while speaking, expresses and thereby anchors his/her biological, psychological or social (e.g. class, region, ethnic group) identity through his/her prosody and gestures. Secondly, it can occur when the body and prosody of the speaker embed what is said in a particular attitude. Thirdly, it can occur when a speaker illustrates what he/she is saying by gesturing something which is similar to what is being talked about. Fourthly, it can occur when a gesture specifies a phrase, e.g. when a pointing gesture specifies what is being referred to by a deictic phrase like *that* or *this*. Fifthly, it can occur as part of communication management, e.g. as part of a speaker change or in giving feedback.

2. *Change of information*: Since addition of information already is a kind of change, what we have in mind here is a kind of modification of information which is not merely addition. An example here might be the use of prosody or facial gestures in a way which suggest an attitude of non-seriousness, irony or satire. Imagine the phrase *he is a nice guy* said with irony. The effect will be almost the same as negation, i.e., "he is not a nice guy". The irony can, however, be more or less integrated in the total message. If it is poorly integrated, we get a kind of double message which might result in what has sometimes been called a "double bind". This can, for example, occur if a parent who wants a teenager to stay home says to the teenager: "Well, you go out and have a good time and I'll stay home and wait for you", while simultaneously with voice quality and facial gestures indicating or displaying disappointment (and resentment).

   A second very different kind of example is provided by gestures used to show that one has made the wrong choice of words (communication management), e.g. *I would like vanilla* (head shake), *chocolate ice cream*.

3. *Reinforcement and support of information*: A very common function of gestures in relation to speech is that of reinforcing and supporting that which is said. This

can be done prosodically using stress, or gesturally by head nods or decisive hand movements.

All three relations discussed above can hold internally between the gestures and the speech of a particular communicator. However, they can also hold interactively, i.e. between different communicators so that the gestures of one communicator add to, change or reinforce the information expressed by another communicator. With the exception of identity-anchoring and own communication management all the functions discussed above in relation to a single communicator would also be applicable to the ways in which the gestures of one communicator can relate to another communicator's contributions in dialog.

6.5 Interaction

In face-to-face communication, each new contribution is usually multimodal, combining vocal verbal with bodily gestural information. In a few cases, however, contributions are unimodal consisting only of vocal verbal or gestural information. Let us now consider the case where a multimodal contribution from one communicator is reacted to by a single- or multimodal contribution from another communicator. The occurrence of the two contributions may from a temporal point of view be either - simultaneous and overlapping or - sequential and non-overlapping.

If we consider the relations from a functional point of view, the following relations seem to be possible:

1. Simultaneous and overlapping contributions

Some of the information which is indicated or displayed by different communicators is overlapping because it is more or less static through an interaction. Examples of this include information concerning identity or physiological state, which can be expressed through clothes or non-changing features of the body. Other information, like emotions and attitudes (e.g. expressed through a sullen or smiling face) can change but often changes slowly, so that one communicator has a good idea of the reactions of the other party as he/she is making his/her contribution. This is also true of some of the ways in which feedback concerning contact, perception, understanding and attention are given through eye gaze and head movements from recipients to the floor-holding communicator while a contribution is being made.

Slightly more active unimodal or multimodal contributions (expressed through vocal words and/or hand or head movements) are often made by recipients to the floor-holding communicator, as he/she is speaking. The function of such overlaps can be of many kinds, but most of them are probably related to "interactive communication management", especially information concerning turn management and feedback. Overlaps can thus be used for turn management, e.g. in attempts to take over the floor (to interrupt). The main use of overlap, however, is probably to provide feedback to the floor-holding communicator about what and how his/her message is being perceived, understood and reacted to. Mostly, this feedback is supportive and involves showing by head-nods and words like *yes* and *mhm* that the message has at least been perceived and understood and that the speaker may therefore continue. In addition to acknowledgement, it often shows acceptance or other attitudes like enthusiasm, disappointment or surprise at what is being said. An interesting special case here is "interactive nodding", i.e. when communicators nod in synchrony throughout several contributions. The speaker nods to reinforce his/her own message and the recipients nod to acknowledge and possibly accept the message.

Overlapping contributions can also be used to give negative feedback showing lack of perception, understanding or acceptance. Looks of puzzlement (using eyebrow raises and/or backwards head-tilts), in combination with question words like *what* can show that the message is not being perceived or understood. Non-acceptance and even rejection can be shown by recipients through satirical smiles, headshakes, sceptical facial gestures and/or negative vocal words.

Even though positive feedback related information is probably the main use of overlapping contributions, occasionally other kinds of information can also be given, often perhaps as extensions of feedback. In this way, a non-floor-holder can add to the floor-holder's message with an illustration or a pointing gesture.

2. Sequential - non-overlapping contributions

If the reaction of one communicator to another is non-overlapping in time, it might still take place in many different positions in relation to the previous contribution, e.g.

- in a pause between the words or constituent phrases of the preceding contribution,
- after the contribution is finished and the communicator is letting go of the floor (turn).

If the contribution takes place in a pause between the words or constituent phrases of a preceding contribution, its functions will, to a large extent, be similar to those that we have discussed above for active simultaneous contributions, i.e., mainly feedback and to some extent turn management. As an example of feedback given sequentially, consider the case of establishing reference in the following example.

A:	Jill's boyfriend		Jack		was here
B:		m		yeah	
		nod		nod	

We see how A and B jointly, step by step establish consensus about who is being referred to. A does this by leaving room for B to signal shared perception and understanding through head nods and feedback words.

The example shows how bodily contributions occurring after a finished contribution function as feedback expressing perception and understanding. However, this position also gives an opportunity to express an attitude toward the point or evocative function of the previous contribution. Thus, a nod can signal agreement after a statement, acceptance of a task after a request and affirmation of a proposition after a yes/no question, in this way providing a kind of context-determined polysemy of the head nod. As already discussed above, such feedback reactions need not be positive, but can also be negative. They can also be extended by providing information in the form of emotional reactions, illustrations or pointing gestures that add to or even change the information provided by the previous contribution.

7. CONCLUSION

The main purpose of this paper has been to show that (and how) body movements are an essential part of interactive "face-to-face" communication, where gestures normally are integrated with speech to form a complex whole which hardly can be understood without considering both gestures and speech and the relation between them.

However, the integration of communicative body movements into a perspective which includes also speech and written communication requires a new understanding of the complex relations which exist between the dimensions of content and the dimensions of expression. This new understanding will include the interplay between, and the integration of, indexical, iconic and symbolic aspects, or to use other similar, commonly used concepts, it will include continuous and discrete, analog and digital aspects of human communication on different levels of awareness and intentionality.

This kind of integration is needed as a counterbalance to the traditional view which has emphasized monologue over dialog, writing over speech, speech over body, symbol over icon, icon over index, discrete over continuous, digital over analog, signal over display and display over indication.

Signalled, digital, discrete, written symbols make up the type of communication where we humans perhaps, in some sense, have made the greatest "artificial" (cultural) contribution. Because of this historical background, writing is the type of communication that has been easiest to study and, if necessary, bring order to by prescriptive means. Since writing is both one of our most important technological social instruments and is fairly open to normative social regulation, writing is also the type of communication which has been most studied.

However, a more complete and correct picture of human communication requires the inclusion of indexical, displayed, analog, continuous, bodily and spoken icons and indices. Expanding scientific description and explanation in this way will most likely not be without problems but will require new ways of thinking of units, relations and operations, both with regard to expression and content. Hopefully, the compensation for this increased degree of difficulty will consist in an increased understanding of human communication not merely as a cultural phenomenon but rather as a phenomenon that has developed as a result of a complex interaction between nature and culture.

8. REFERENCES

Ahlsén, Elisabeth. The Nonverbal Communication of Aphasics in Conversation. *Gothenburg Papers in Theoretical Linguistics 48.* Göteborg University, 1985.

Ahlsén, Elisabeth. Body Communication and Speech in a Wernicke's Aphasic - A Longitudinal Study. *Journal of Communication Disorders* 24, 1-12, 1991.

Allwood, Jens. Linguistic Communication as Action and Cooperation. *Gothenburg Monographs in Linguistics 2.* Göteborg University, 1976.

Allwood, Jens. Ickeverbal kommunikation - en översikt. In: Stedje, Astrid & af Trampe, Peter (Eds). *Tvåspråkighet.* Stockholm: Akademilitteratur, 1979.

Allwood, Jens. En analys av kommunikation. In: Nowak, Andrén & Strand (Eds). *Kommunikationsprocesser.* Mass No. 7, Centrum för Masskom. forskning. Stockholm University, 1983.

Allwood, Jens. Om det svenska systemet för språklig återkoppling. In: Linell, P., Adelswärd, V., Nilsson, T., & Pettersson, P.A. (Eds). *Svenskans Beskrivning 16.1* (SIC 21a), Linköping University, Tema Kommunikation, 1988.

Allwood, Jens, Joakim Nivre & Elisabeth Ahlsén. Speech Management: On the Non-Written Life of Speech. *Nordic Journal of Linguistics,* 13, 3-48, 1990.

Allwood, Jens, Joakim Nivre & Elisabeth Ahlsén. On the Semantics and Pragmatics of Linguistic Feedback. *The Journal of Semantics,* 9.1, 1992.

Allwood, Jens. Cooperation and Flexibility in Multimodal Communication. Forthcoming in Beun, R.-J., & Bunt, Harry (Eds). *Proceedings of the Second International Conference on Cooperative Multimodal Communication. CMC/98* Springer Verlag. Lecture Notes on AI series, 2000.

Argyle, Michael. *Bodily Communication,* London: Methuen, 1975.

Austin, John Langshaw. *How to do Things with Words.* Oxford University Press, 1962.

Cassell, J., J. Sullivan, S. Prevost & E. Churchill. *Embodied Conversational Agents.* MIT Press: Cambridge, Mass.

Davis, Martha. *Interaction Rhythms.* New York: Human Sciences Press, 1982.

Duncan, Starkey and Donald Fiske. *Face-to-Face Interaction.* Lawrence Erlbaum. Hillsdale. N.J., 1977.

Ekman, Paul & Wallace Friesen. The Repertoire of Nonverbal Behavior: Categories, Origins, Usage and Coding. *Semiotica,* 1: 49-98, 1969.

Fast, Julius. *Kroppsspråket.* (orig. Body Language, N.Y.), 1973.

Fex, B. & A.-C. Månsson. The Use of Gestures as a Compensatory Strategy in Adults with Acquired Aphasia Compared to Children with Specific Language Impairment (SLI). *Journal of Neurolinguistics,* Vol. 11, Nr 1/2, 191-206, 1998.

Hirsch, Richard. *Argumentation, Information and Interaction.* Gothenburg Monographs in Linguistics 7. Göteborg University, 1989.

Hjortsjö, Carl Herman. *Människans Ansikte och Mimiska Språket.* Studentlitteratur, Malmö, 1969.

Johansson, Gunnar. Visual Perception of Biological Motion and a Model for its Analysis. *Perception and Psychophysics,* 14, 201-211, 1973.

Kendon, Adam. *Nonverbal Communication. Interaction and Gesture.* The Hague: Mouton, 1981.

Key, Mary Ritchie. *Nonverbal Communication Today.* Berlin: Mouton, 1982.

Knapp, Mark. *Nonverbal Communication in Human Interaction.* New York: Holt, Rinehart & Winston. 1978 and later editions.

Lipset, David. *Gregory Bateson - the Legacy of a Scientist.* Englewood Cliffs. N.J.: Prentice Hall, 1980.

McNeill, David. *The Conceptual Basis of Language.* Hillsdale: Lawrence Erlbaum, 1979.

Morris, Desmond. *Manwatching.* Oxford: Elsevier, 1977.

Nivre, Joakim. *Situations, Meaning and Communications. A Situation Theoretic Approach To Meaning in Language and Communication.* Gothenburg Monographs in Linguistics 11. Göteborg Univ, 1992.

Øyslebø Olaf. *Ickeverbal kommunikasjon.* Oslo: Universitetsforlaget, 1981.

Peirce, Charles Sanders (1902) in Buchler, J. *Philosophical Writings of Pierce.* New York: Cover, 1955.

Sacks, Harvey, Emanuel Schegloff, & Gail Jeffersson. A Simplest Systematics for the Organization of Turntaking in Conversation. *Language,* 50: 696-735, 1974.

Schiffer, Stephen. *Meaning.* Oxford University Press, 1972.

9. AFFILIATION

Jens Allwood, Department of Linguistics, Göteborg University, Box 200, SE 405 30 Göteborg, Sweden

DAVID MCNEILL, FRANCIS QUEK,
KARL-ERIK MCCULLOUGH, SUSAN DUNCAN,
ROBERT BRYLL, XIN-FENG MA,
AND RASHID ANSARI

DYNAMIC IMAGERY IN SPEECH AND GESTURE

1. INTRODUCTION

Someone begins to describe an event and almost immediately her hands start to fly. The movements seem involuntary and indeed unconscious, yet they take place vigorously and abundantly. Why is this happening? Whatever the reason, our person is not alone. Popular beliefs notwithstanding, every culture produces gestures. Gesturing is a phenomenon that passes almost without notice but it is omnipresent. If you watch someone speaking, in almost any language, and under nearly all circumstances, you will see what appears to be a compulsion to move the head, hands and arms in conjunction with speech. Speech, we know, is the actuality of language. But what are these gestures? They are not compensations for missing words or inarticulate speech – if anything, gestures are positively related to fluency and complexity of speech – the more articulate the speech, the more gesture.

Such gestures have been called gesticulations by Kendon (1988). They need to be distinguished from culturally codified gestures such as the 'OK' sign or waving goodbye, known as emblems. Gesticulations are characterized by an obligatory accompaniment of speech, a lack of language-defining properties, idiosyncratic form-meaning pairings, and a precise synchronization of meaning presentations in gestures with co-expressive speech segments. This stable of characteristics shows that gestures and speech are systematically organized in relation to one another. The gestures are meaningful. They form meaningful, non-redundant combinations with the speech segments with which they synchronize, despite the fact that they are idiosyncratic and ephemeral. Language actually includes motion of the limbs along with the usual linguistic components - words, phrases, etc.

The argument of this paper is that these gestures are part of our thinking process, processes that take place automatically as the mind engages itself with language (cf. McNeill, 1992; McNeill & Duncan, 2000). Not any thinking process but thinking that emerges with and is evoked through language. Dan Slobin in 1987 introduced a technical term for this cognitive mode – "thinking for speaking". Gestures are irresistible because thinking for speaking is irresistible.

B. Granström et al. (eds.), Multimodality in Language and Speech Systems, 27–44.

Our goal is to elaborate this argument and explore its plausibility and depth, making use of new techniques of motion analysis.

To illustrate the kind of everyday gesture that we mean, consider the following example. A speaker recounting a cartoon story lifts her hand upward with the meaning that a character is climbing up. Her gesture is not a codified cultural gesture such as the 'OK' sign. It is imagery created on the fly at the moment of speaking and is part of the speaker's meaning at this moment. The rising hand embodies upwardness and shares this meaning with the most co-expressive parts of the accompanying utterance, "[and he climbs **up the pipe**]."[1] By 'image' we mean a semiotic object with the following crucial property: the meanings of the parts are determined by the meaning of the whole. This is called the global property (McNeill, 1992). The hand is a character, it is not a hand, because it is part of a gesture whose overall meaning is a character rising upward. The global property is the opposite of the compositional property found in linguistic objects. A gesture is top-down while a sentence is bottom-up. A sentence is composed out of independently meaningful parts, words or morphemes. The meaning of the word 'hand' does not change according to the overall meaning of the sentence. The meaning of the hand in a gesture does change depending on the overall meaning of the gesture and can mean a hand or something else altogether in another gesture. The temporal alignment of imagery and speech is crucial, for it suggests that upness is being thought of in *two cognitive frameworks simultaneously*. One framework is instantaneous, visuospatial and actional, and is characterized by an imagistic mode of cognition visible in the gesture. The other is linear, segmented and language-like, and is actualized in the speech. These modes are *simultaneous*; it is not that one leads to the other. The image of upness was co-present with the segments "up" and "the pipe." Thinking takes place in both frameworks at once. Simultaneity is important because it suggests constraints on the mechanisms of speaking and thinking beyond the obvious requirements of communication.

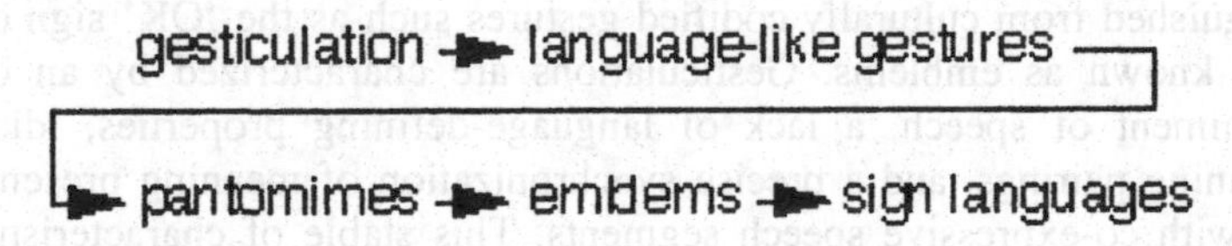

Figure. Kendon's continuum of gestures (reproduced from Hand and Mind, McNeill, 1992).

A key question is the source of constraint on gesture form and timing. Kendon (1988) described a taxonomy of gesture summarized and organized as a continuum in Figure 1. At one end of the continuum is *gesticulation*, which describes the free-form gesturing which typically accompany verbal discourse. At the other end of this continuum are the sign languages of the deaf (such as American Sign Language, or ASL), which are characterized by complete lexical and grammatical specification. In between, we have 'language-like gestures,' in which the speaker inserts a gesture in the place of a syntactic unit during speech; 'pantomimes,' in which the gesturer creates an iconic and motion facsimile of the referent; and 'emblems,' which are codified gestural expressions that are not governed by any formal grammar (such as OK or the 'thumbs-up' gesture to signify 'all is well'). This paper focuses on the 'gesticulation' end of this continuum.

We learn from gesture that when the mind engages language it also engages instantaneous imagery as an integral part. Gesture and speech are not subservient to one another. One is not an afterthought to enrich or augment the other. Instead, they proceed together from the same 'idea units'. Contrary to a text-based tradition that has dominated linguistics and psycholinguistics from their historical beginnings, language contains analogic, imagistic elements that are as defining of language capacity as the time-honored analytic elements of words, phrases, clauses, sentences, and texts. The postulate that language and imagery are inseparable explains the power of gestures to open a window onto the mind and to reveal cross-linguistic differences in thinking for speaking. This is an insight that cannot be deduced from text-based data and requires theoretical concepts of language in new domains.

With gesture, we are invited inside the mental life of another person, not only because gesture offers 'evidence' of mind, but because the gesture, as such, is one of the very aspects of the mind. Some imagery is purely mental and without any apparent bodily enactment, but with gesture the image receives a material existence and gains thereby in actuality as well as observability. To put these points together – that language contains imagery as an integral part, and that gesture is the most developed form of it – we say that gesture is part of the actualization of language, along with speech itself, and is no less a crucial part of this actualization.

We seek an understanding of gesture and speech in natural human conversation. We believe that an understanding of the constants and principles of speech-gesture cohesion is essential to its application in human computer interaction involving both modalities. Several observations are necessary to frame the premise of such conversational interaction. While gesture may be extended to include head and eye gestures, facial gestures and body motion, we shall explore here only the relationship of hand gestures and speech. Conversational gestures differ from manipulative movement in several significant ways. First, because the intent of the latter is for manipulation, there is no guarantee that the salient features of the hands are visible. Second, the dynamics of hand movement in manipulative gestures differ significantly from conversational gestures. Third, manipulative gestures may typically be aided by tactile and force feedback from the object (virtual or real) being manipulated, while conversational gestures are typically performed without such constraints.

We shall present an underlying psycholinguistic model by which gesture and speech entities may be integrated, describe our research method that encompasses processing of the video data and psycholinguistic analysis of the underlying discourse, and present the results of our analysis. Our purpose, in this paper, is to motivate a perspective of conversation interaction, and to show that the cues for such interaction are accessible in video data. In this study, our data comes from a single video camera, and we consider only gestural motions in the camera's image plane.

2. PSYCHOLINGUISTIC BASIS

Gesture and speech clearly belong to different modalities of expression but they are linked on several levels and work together to present the same semantic idea units. The two modalities are not redundant; they are 'co-expressive,' meaning that they arise from a shared semantic source but are able to express different aspects of it, overlapping this source in their own ways. A simple example will illustrate. In the living space text we present below, the speaker describes at one point entering a house with the clause, "when you open the doors." At the same time she performs a two-handed anti-symmetric gesture in which her hands, upright and palms facing forward, move left to right several times. Gesture and speech arise from the same semantic source but are non-redundant; each modality expresses its own part of the shared constellation. Speech describes an action performed in relation to it, gesture shows the shape and extent of the doors and that there are two of them rather than one; thus speech and gesture are co-expressive. Since gesture and speech proceed from the same semantic source, one might expect that the semantic structure of the resulting discourse to be accessible through both the gestural and speech channels.

2.1. The Catchment Concept

The 'catchment' concept provides a locus along which gestural entities may be viewed to provide access to the discourse structure. A catchment is a term for discourse units inferred on the basis of gesture information. Catchments are recognized when gesture features recur in at least two (not necessarily consecutive) gestures. The logic is that imagery generates the gesture features; recurrent imagery suggests a common discourse theme. A catchment is a kind of thread of consistent visuospatial imagery running through a discourse segment. The catchment is a kind of thread of visuospatial imagery through a discourse that reveals the separate parts cohering into larger discourse units. By discovering a given speaker's catchments, we can see what for this speaker goes together into larger discourse units – what meanings are seen as similar or related and grouped together, and what meanings are isolated and thus seen by the speaker as having distinct or less related meanings. Consider one of the most basic gesture features, handedness.

Gestures can be made with one hand (1H) or two (2H); if 1H, they can be made with the left hand (LH) or the right (RH); if 2H, the hands can move and/or be positioned in mirror images or with one hand taking an active role and the other a more passive 'platform' role. Noting groups of gestures that have the same values of handedness can identify catchments. We can add other features such as shape, location in space, and trajectory (curved, straight, spiral, etc.), and consider all of these as also defining possible catchments. A given catchment could, for example, be defined by the recurrent use of the same trajectory and space with variations of hand shapes. This would suggest a larger discourse unit within which meanings are contrasted. Individuals differ in how they link up the world into related and unrelated components, and catchments give us a way of detecting these individual characteristics or cognitive styles.

3. EXPERIMENTAL METHOD

Hand gestures are seen in abundance when people describe spatially organized information. In our gesture and speech elicitation experiment, subjects are asked to describe their living quarters to an interlocutor. This conversation is recorded on a Hi-8 tape. Figure 2 is a frame from the experimental sequence that is presented here. Two independent sets of analyses are performed on the video and audio data. The first set entails the processing of the video data to obtain the motion traces of both of

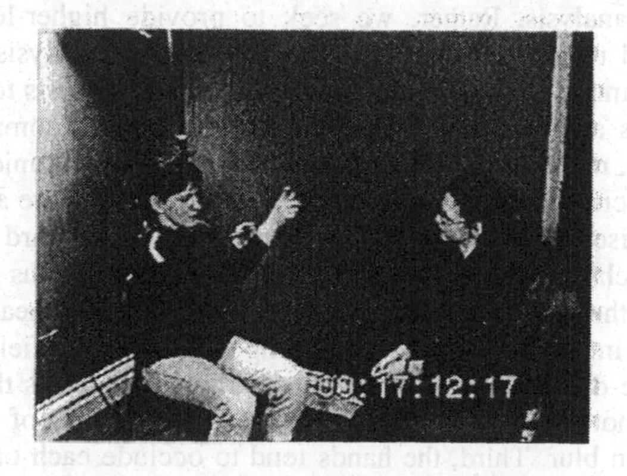

Figure 2. A frame of the video data collected in our gesture elicitation experiment.

the subject's hands. The synchronized audio data are also analyzed to extract the fundamental frequency signal and speech power amplitude (in terms of the RMS value of the audio signal). The second set of analyses entails expert transcription of the speech and gesture data. This transcription is done by carefully transcribing and analyzing the Hi-8 video tape using a frame-accurate video player to correlate the speech with the gestural entities. We also perform a higher-level analysis using the transcribed text alone. Finally, the results of the psycholinguistic analyses are compared against the features computed in the video and audio data. The purpose of this comparison is to identify the cues accessible in the gestural and audio data that correlate well with the expert psycholinguistic analysis. We shall discuss each step in turn.

3.1. Extraction of Hand Motion Traces for Monocular Video

The overarching goal of our technical work is to build data and computational bridges among the disciplines represented in this paper – linguistics, psycholinguistics, and machine vision. The research proceeds along two paths toward two sets of complementary objectives. First, we research and develop the enabling technology for providing computation and access to the audio/video data and the associated derived information that support the needs of the science project. This includes the identification of the primitives that have to be computed, and the representation and organization of these primitives in close cooperation with all participating scientists. Second, the participating science project pursues its objectives around the locus of integrated verbal and nonverbal communication. These objectives have informed the development of the computational bridge and are, in turn, enhanced by use of the resulting technology. By directly computing and representing the free-flow nature of speech prosody and motion velocities and accelerations, we can provide researchers with a new level of quantification of combined speech and motion dynamics. Our goal has not been to do away with expert perceptual analysis. Rather, we seek to provide higher-level objects and representations and to mitigate the labor-intensiveness of analysis that has access only to the time stamp of the video signal. We automate analysis to extract discrete audio/video chunks appropriate to the participating research domains and provide interfaces to access, manipulate, and analyze multi-modal communicative entities.

In the work described here our purpose is to see what cues are afforded by gross hand motion for discourse structuring. Hand gestures in standard video data pose several processing challenges. First, one cannot assume contiguous motion. A sweep of the hand across the body can span just 0.25 s. to 0.5 s. This means that the entire motion is captured in 7 to 15 frames. Depending on the camera field-of-view on the subject, inter-frame displacement can be quite large. This means that dense optical flow methods cannot be used. Second, because of the speed of motion, there is considerable motion blur. Third, the hands tend to occlude each other. We apply a parallelizable fuzzy image processing approach known as *Vector Coherence Mapping* (VCM) (Quek et al, 1999; Quek & Bryll, 1998) to track the hand motion. VCM is able to apply spatial coherence, momentum (temporal coherence), motion, and skin color constraints in the vector field computation by using a fuzzy-combination strategy, and produces good results for hand gesture tracking. Figure 3 illustrates how VCM applies a spatial coherence constraint (minimizing the directional variance) in vector field computation. Assume three feature points p at time t (represented by the squares at the top of the figure) moves to their new locations (represented by circles) in the next frame. If all three feature points correspond equally to each other, an application of convolution to detect matches (e.g. by a *Absolute Difference Correlation*, ADC) from each p would yield correlation maps with three hotspots (shown as N3 in the middle of Figure 3). If all three correlation maps are normalized and summed, we obtain the vector coherence map (*vcm*) at the bottom of the figure, the 'correct' correlations would reinforce each other, while the chance correlations would not. Hence a simple weighted summation of neighboring *vcm*'s would yield a vector that minimizes the local variance in the computed vector field. A normalized *vcm* computed for each feature point can be thought of as a 'likelihood' map for the spatial variance-minimizing vector at that point. This may be used in a 'fuzzy image processing' process where other constraints may be 'fuzzy-ANDed' with it. We use a temporal constraint based on momentum, and a skin-color constraint to extract the required hand-motion vector fields. Figure 4 shows the effect of incorporating the skin-color constraint to clean up the computed vector fields. See Quek et al (1999) and Quek & Bryll (1998) for detailed derivation of the approach.

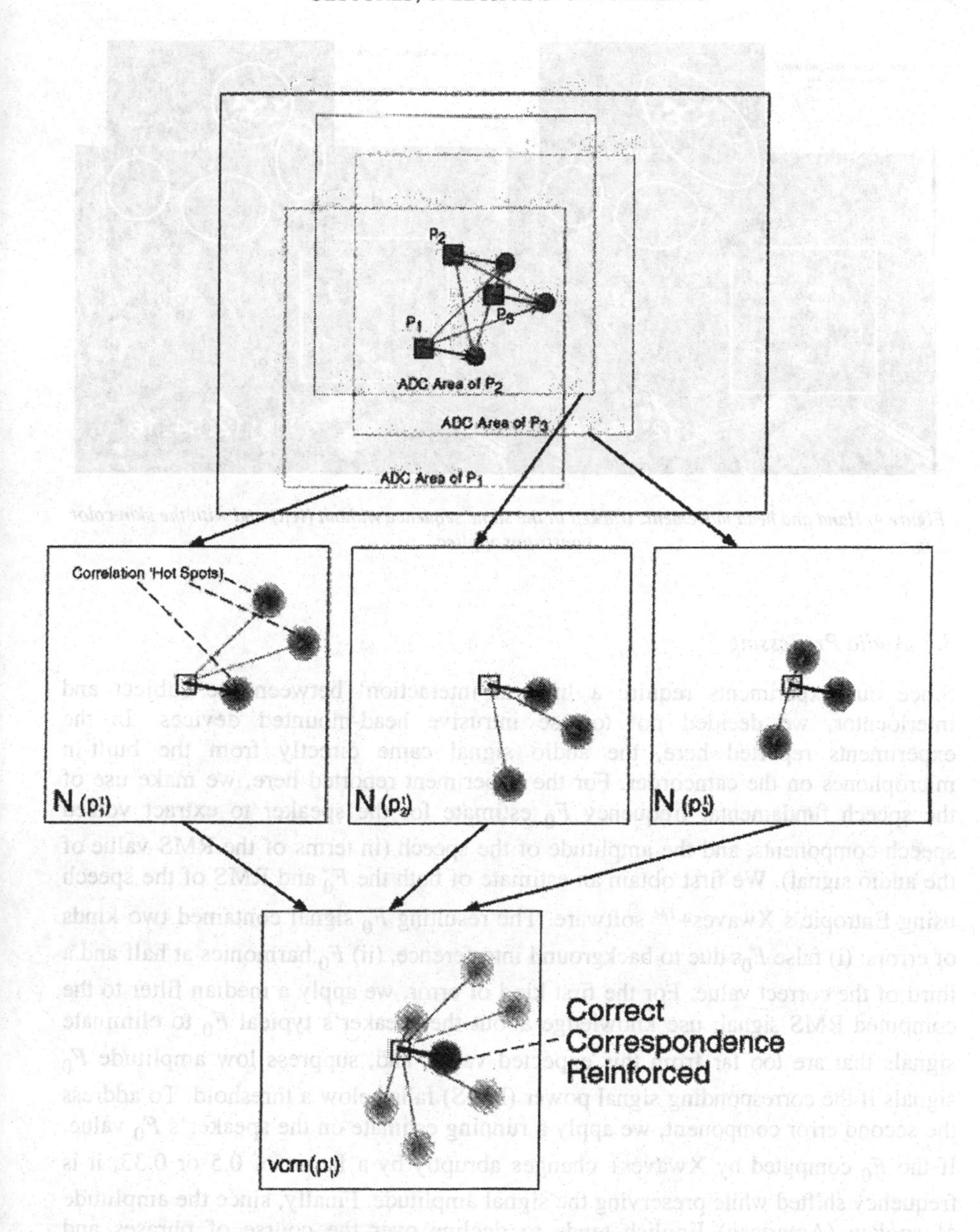

Figure 3. Spatial Coherence Constraint in VCM.

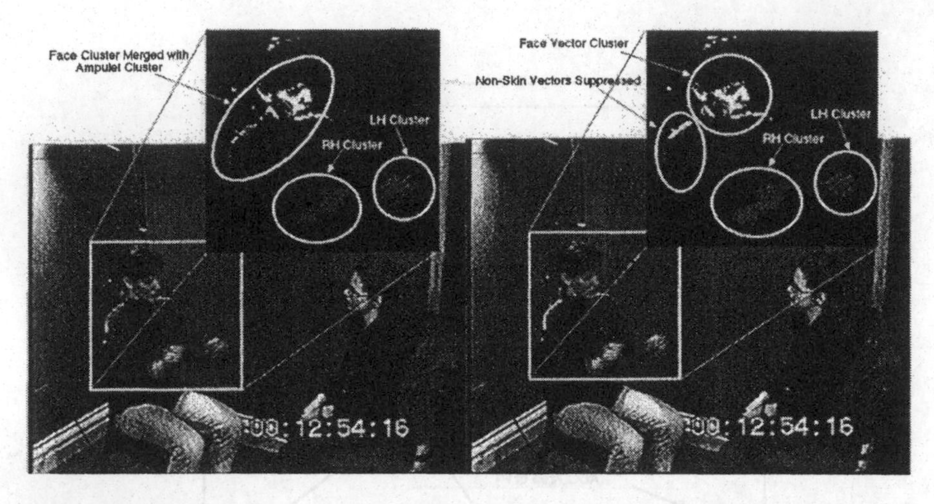

Figure 4. Hand and head movements tracked in the same sequence without (left) and with the skin color constraint applied.

3.2. Audio Processing

Since our experiments require a 'natural interaction' between the subject and interlocutor, we decided not to use intrusive head-mounted devices. In the experiments reported here, the audio signal came directly from the built-in microphones on the camcorder. For the experiment reported here, we make use of the speech fundamental frequency F_0 estimate for the speaker to extract voiced speech components, and the amplitude of the speech (in terms of the RMS value of the audio signal). We first obtain an estimate of both the F_0 and RMS of the speech using Entropic's Xwaves+^{TM} software. The resulting F_0 signal contained two kinds of errors: (i) false F_0s due to background interference, (ii) F_0 harmonics at half and a third of the correct value. For the first kind of error, we apply a median filter to the computed RMS signal; use knowledge about the speaker's typical F_0 to eliminate signals that are too far from this expected value; and, suppress low amplitude F_0 signals if the corresponding signal power (RMS) falls below a threshold. To address the second error component, we apply a running estimate on the speaker's F_0 value. If the F_0 computed by Xwaves+ changes abruptly by a factor of 0.5 or 0.33, it is frequency shifted while preserving the signal amplitude. Finally, since the amplitude of spoken (American) English tends to decline over the course of phrases and utterances (Ladd, 1996), we apply a linear declination estimator to group F_0 signal intervals into such declination units. We compute the pause durations within the F_0 signal. The remaining F_0 signal is mapped onto a parametric trend function to locate likely phrase units. For further discussion, see Ansari et al. (1999).

3.3. Psycholinguistic Analysis

Perceptual analysis of video (analysis by unaided ear and eye) plays an important role in such disciplines as psychology, psycholinguistics, linguistics, anthropology, and neurology. In psycholinguistic analysis of gesture and speech, researchers microanalyze videos of subjects using a high quality video-cassette recorder that has a digital freeze capability down to the specific frame. The analysis typically proceeds in three iterations. First, the speech is carefully transcribed by hand, and then typed into a text document. The beginning of each linguistic unit (typically a phrase) is marked by the time-stamp of the beginning of the unit in the video tape. Second, the researcher revisits the video and annotates the text, marking co-occurrences of speech and gestural phases (rest-holds, pre-stroke and post-stroke holds, gross hand shape, trajectory of motion, gestural stroke characteristics, etc.). The researcher also inserts locations of audible breath pauses, speech disfluencies, and other salient comments. Third, all these data are formatted onto a final transcript for psycholinguistic analysis. This is a painstaking process that requires a week to ten days to analyze about a minute of discourse. The outcome of the psycholinguistic analysis process is a set of detailed transcripts. We have reproduced the first page of the transcript for our example data set in Figure 5. The gestural phases are annotated and correlated with the text (by under-lining, bolding, brackets, symbols insertion etc.), the F_0 units (numbers above the text), and the video-tape time stamp (time signatures to the left). Comments about gesture details are placed under each transcribed phrase. In addition to this, we perform a 'levels of discourse' analysis based on the transcribed text alone without looking at the video (see below).

3.4. Comparative Analysis

In our experiments we analyze the gestural and audio signals in parallel with the perceptual psycholinguistic analysis to find cues for high level discourse segmentation that are accessible to computation. The voiced units in the F_0 plots are numbered and related to the text manually. These are plotted together with the gesture traces computed using VCM. First, we perform a perceptual analysis of the data to find features that correlate well between the gesture signal and the high-level speech content. We call this the discovery phase of the analysis. Next, we devise algorithms to extract these features. We shall present the features and cues we have found in our data.

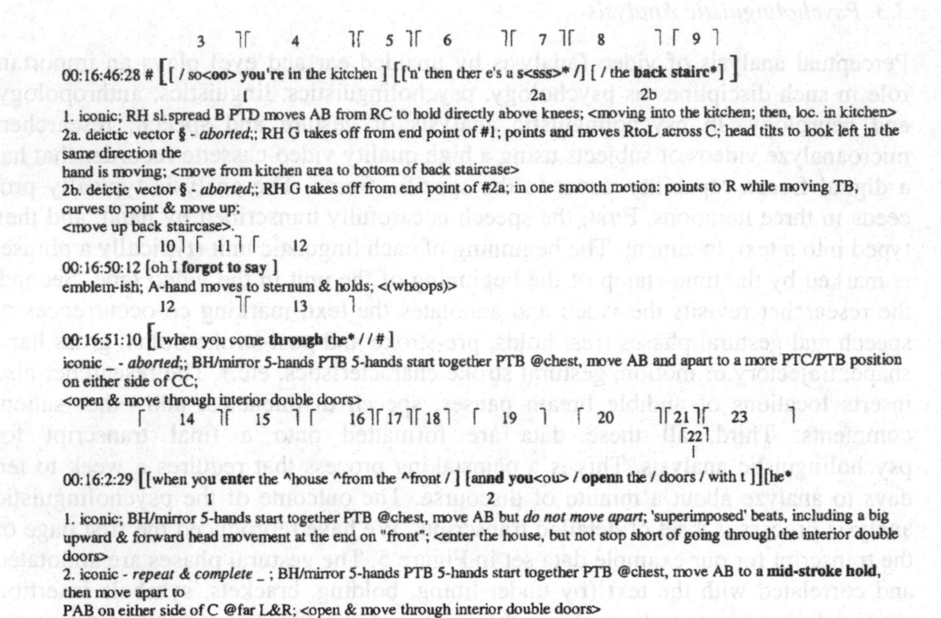

3 ⌉⌈ 4 ⌉⌈ 5 ⌉⌈ 6 ⌉⌈ 7 ⌉⌈ 8 ⌉⌈ 9 ⌉

00:16:46:28 # [[/ so<oo> **you're** in the kitchen] [['n' then ther e's a **s<sss>*** /] [/ the **back stairc***]]

1 2a 2b

1. iconic; RH sl.spread B PTD moves AB from RC to hold directly above knees; <moving into the kitchen; then, loc. in kitchen>
2a. deictic vector § - *aborted*;; RH G takes off from end point of #1; points and moves RtoL across C; head tilts to look left in the same direction the
hand is moving; <move from kitchen area to bottom of back staircase>
2b. deictic vector § - *aborted*;; RH G takes off from end point of #2a; in one smooth motion: points to R while moving TB, curves to point & move up;
<move up back staircase>.

⌈ 10 ⌉⌈ 11 ⌉⌈ 12

00:16:50:12 [oh I **forgot to say**]
emblem-ish; A-hand moves to sternum & holds; <(whoops)>

12 ⌉⌈ 13 ⌉

00:16:51:10 [[when you come **through** the* / / #]
iconic _ - *aborted*;; BH/mirror 5-hands PTB 5-hands start together PTB @chest, move AB and apart to a more PTC/PTB position on either side of CC;
<open & move through interior double doors>

⌈ 14 ⌉⌈ 15 ⌉⌈ 16 ⌉⌈ 17 ⌉⌈ 18 ⌉ ⌈ 19 ⌉ ⌈ 20 ⌉⌈ 21 ⌉⌈ 23 ⌉
⌈22⌉

00:16:2:29 [[when you **enter** the ^house ^from the ^front /] [**annd you**<ou> / **openn** the / doors / with t]][he*

1 2

1. iconic; BH/mirror 5-hands start together PTB @chest, move AB but *do not move apart*; 'superimposed' beats, including a big upward & forward head movement at the end on "front"; <enter the house, but not stop short of going through the interior double doors>
2. iconic - *repeat & complete* _ ; BH/mirror 5-hands PTB 5-hands start together PTB @chest, move AB to a **mid-stroke hold**, then move apart to
PAB on either side of C @far L&R; <open & move through interior double doors>

Figure 5: Sample analysis transcript

4. RESULTS

Figures 6 and 7 are summary plots of 32 seconds of experimental data plotted with the video frame number as the time scale (480 frames in each chart to give a total of 960 frames at 30 fps). The x-axis of these graphs is the frame numbers. The top two plots of each figure describe the horizontal (x) and vertical (y) hand positions, respectively. The horizontal bars under the y direction plot is an analysis of hand motion features – LH hold, RH hold, 2H anti-symmetric, 2H symmetric (mirror symmetry), 2H (no detected symmetry), single LH and single RH, respectively. Beneath this is the fundamental frequency $F0$ plot of the audio signal. The voiced utterances in the $F0$ plot are numbered sequentially to facilitate correlation with the speech transcript. We have reproduced the synchronized speech transcript at the bottom of each chart, as they correlate with the $F0$ units. The vertical shaded bars that run across the charts mark the durations in which both hands are determined to be stationary (holding). We have annotated the hand motion plots with parenthesized markers and brief descriptive labels. We will use these labels along with the related frame-number durations for our ensuing discussion.

4.1. Discourse Segments

4.1.1. Linguistic Segmentation.

Barbara Grosz and colleagues (Nakatani et al, 1995) have devised a systematic procedure for recovering the discourse structure from a transcribed text. The method consists of a set of questions with which to guide analysis and uncover the speaker's goals in producing each successive line of text. Following this procedure with the text displayed in Figures 6 and 7 produces a four-level discourse structure. This

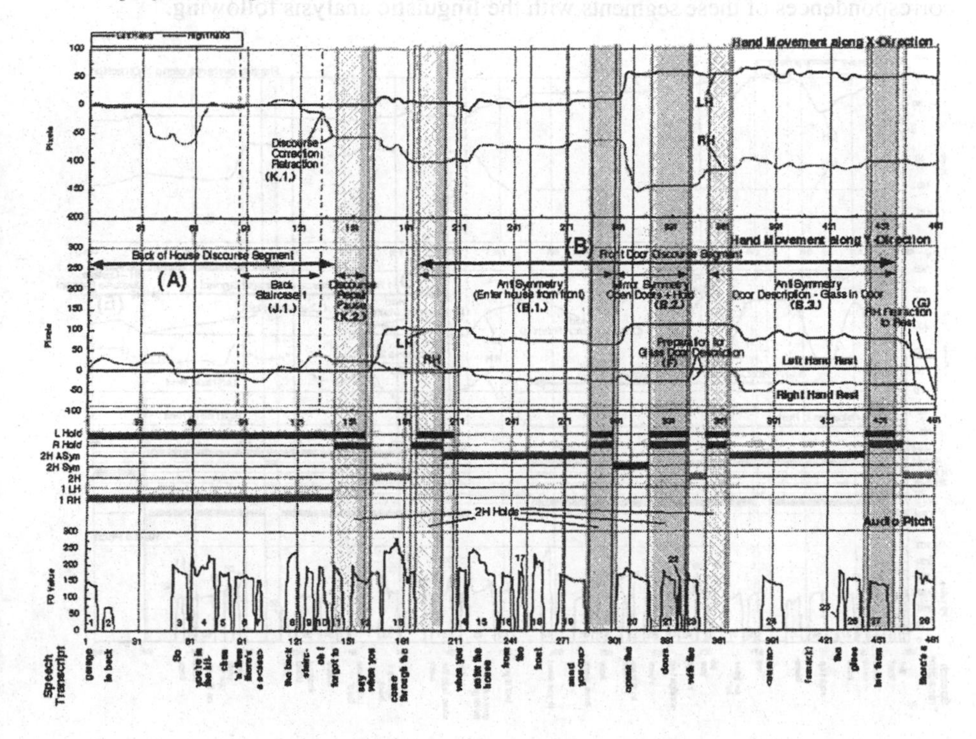

Figure 6. Hand position, analysis and F0 graphs for frames 1-480.

structure can be compared to the discourse segments inferred from the objective motion patterns shown in the gestures. The gesture-based and text-based pictures of the discourse segmentation are independently derived but the resulting correlation is remarkably strong, implying that the gesture features were not generated haphazardly but arose as part of a structured, multi-level process of discourse-building. Every gesture-feature corresponds to a text-based segment. Discrepancies arise where the gesture structure suggests discourse segments that the text-based hierarchy fails to reveal, implying that the gesture modality has captured additional discourse segments that did not find their way into the textual transcription. The uppermost level of the Grosz-type hierarchy can be labeled "Locating the Back

Staircase." It is at this level that the discourse as a whole had its significance. The middle level concerned the first staircase and its location, and the lowest level the front of the house and the restarting of the house tour from there.

4.1.2. Gesture-Speech Discourse Correlations.

Labels (A) through (E) mark the discourse segments accessible from the gestural traces independently from the speech data. These segments are determined solely from the hold and motion patterns of the speaker's hands. We summarize the correspondences of these segments with the linguistic analysis following.

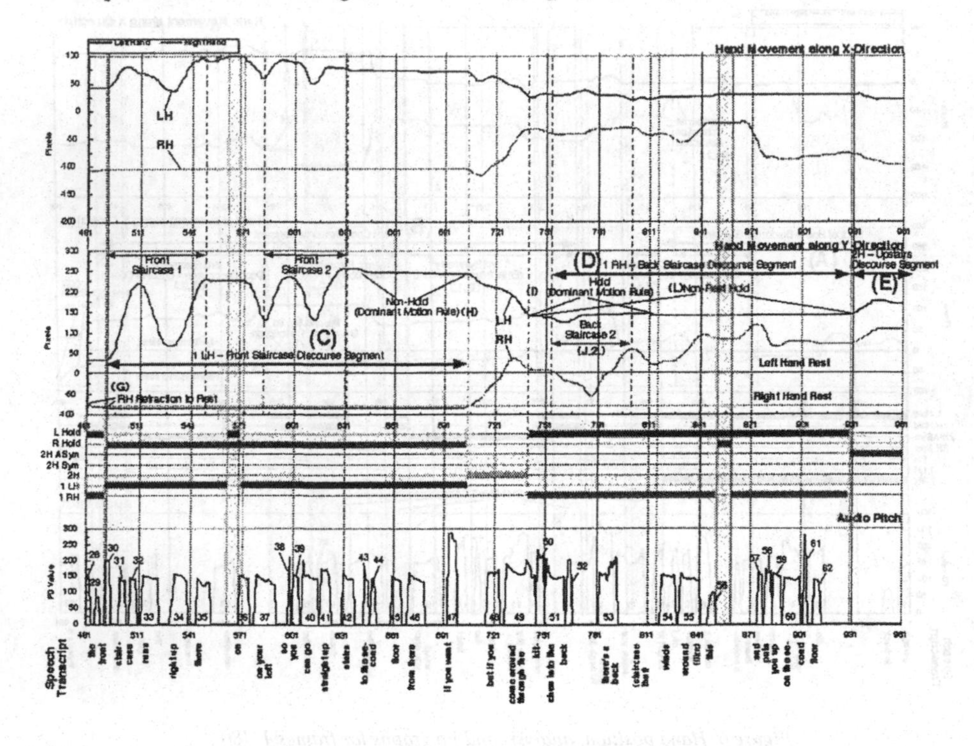

Figure 7. Hand position, analysis and F_0 graphs for frames 481-961.

(A) *Back-of-house discourse segment, 1 RH (Frames 1–140):* These one-handed gestures, all with the RH, accompany the references to the back of the house that launch the discourse. This 1H catchment is replaced by a series of 2H gestures in (B), marking the shift to a different discourse purpose, that of describing the front of the house. Notice this catchment feature of 1H-RH gestures (i.e. the LH is holding) reprises itself in segment (D) when the subject returns to describing the back of the house.

(B) *Front door discourse segment, 2 Synchronized Hands (Frames 188–455):* Two-handed gestures occur when the discourse theme is the front of the house, but there are several variants and these mark sub-parts of the theme – the existence of the front door, opening it, and describing it. Each sub-theme is initiated by a gesture hold, marking off in gesture the internal divisions of the discourse hierarchy. These sub-divisions are not evident in the text and thus not picked up by the text-only analysis that produced the purpose hierarchy and its segmentation (described in Section 4.1.1). This finer grained segmentation is confirmed by psycholinguistic analysis of the original video.

(B.1.) *'Enter house from front' discourse segment 2H Anti-symmetric (Frames 188–298):* Anti-symmetric two-handed movements iconically embody the image of the two front doors; the anti-symmetric movements themselves contrast with the following mirror-image movements, and convey, not motion as such, but the surface and orientation of the doors.

(B.2.) *'Open doors' discourse segment 2H Mirror Symmetry (Frames 299–338):* In contrast, with the preceding two-handed segment, this gesture shows opening the doors and the hands moving apart. This segment terminates in a non-rest two-handed hold of sizeable duration of more than 0.75 s (all other pre-stroke and post-stroke holds are less than 0.5 s in duration). This suggests that it is itself a 'hold-stroke' (i.e. an information-laden component). This corresponds well with the text transcription. The thrusting open of the hands indicating the action of opening the doors (coinciding with the words: 'open the'), and the 2H hold-stroke indicating the object of interest (coinciding with the word: 'doors'). This agrees with the discourse analysis that carries the thread of the 'front door' as the element of focus.

Furthermore, the 2H mirror symmetric motion for opening the doors carries the added information that these are double doors (information unavailable from the text transcription alone).

(B.3.) *Door description discourse segment 2H Anti-symmetric (Frames 351–458):* The form of the doors return as a sub-theme in their own right, and again the movement is anti-symmetric, in the plane of the closed doors.

(C) *Front staircase discourse segment, 1 LH (Frames 491–704):* The LH becomes active in a series of distinctive up-down movements coinciding exactly with the discourse goal of introducing the front staircase. These are clearly one-handed gestures with the RH at the rest position.

(D) *Back staircase discourse segment 1 RH (Frames 754–929):* The gestures for the back staircase are again made with the RH, but now, in contrast to the (A) catchment, the RH is at a non-rest hold, and still in play from (C). This changes in the final segment of the discourse.

(E) *'Upstairs' discourse segment 2H synchronized (Frames 930–):* The LH and RH join forces in a final gesture depicting ascent to the second floor via the back staircase. This is another place where gesture reveals a discourse element not recoverable from the text (no text accompanied the gesture).

4.2 Other Gestural Features

Beside the overall gesture hold analysis, this 32 seconds of discourse also contains several examples of gestural features and cues. We have labeled these (F) through (L). The following discussion summarizes the features we found under these labels.

(F) *Preparation for glass door description (Frames 340–359):* In the middle of the discourse segment on the front door (B), we have the interval marked (F) which appears to break the symmetry. This break is actually the preparation phase of the RH to the non-rest hold (for both hands) section that continues into the strongly anti-symmetric (B.3.) 'glass door' segment. This clarifies the interpretation of the 2H holds preceding and following (F). The former is the post-stroke hold for the 'open doors' segment (B.2.), and latter is the pre-stroke hold for segment (B.3.). Furthermore, we can then extend segment (B.3.) backward to the beginning of (F). This would group *F*0 unit 23 ('with the') with (B.3.). This matches the discourse segmentation 'with the ... <uumm> glass in them'. It is significant to note that the subject has interrupted her speech stream and is 'searching' for the next words to describe what she wants to say. The cohesion of the phrase would be lost in a pure speech pause analysis. We introduce the rule for gesture segment extension to include the last movement to the pre-stroke hold for the segment.

(G) *RH retraction to rest (Frames 468–490 straddles both charts):* The RH movement labeled (G) spanning both plots terminate in the resting position for the hand. We can therefore extend the starting point of the target rest backward to the start of this retraction for discourse segmentation. Hence, we might actually begin the (C) front staircase discourse segment marked by the 1 LH feature backward from frame 491 to frame 340. This matches the discourse analysis from *F*0 units 28–30: ".. . there's the front- . . . " This provides us with the rule of rest-extension to the start of the last motion that results in the rest. The pauseless voiced speech section from *F*0 units 28–35 would provide complementary evidence for this segmentation. We have another example for this rule in LH the motion preceding the hold labeled (I), effectively extending the 'back staircase' discourse segment (D) backward to frame 705. Again this corresponds well with the speech transcript.

(H) & (I) *Non-Hold for (H) in (C) (Frames 643–704) and Hold for (I) in (D) (Frames 740–811):* The LH was judged by the expert coder to be not holding in (H) while it was judged to be holding in (I). An examination of the video shows that in (H) the speaker was making a series of small oscillatory motions (patting motion with her wrist to signify the floor of the 'second floor') with a general downward trend. In segment (I), the LH was holding, but the entire body was moving slightly because of the rapid and large movements of the RH. This distinction cannot be made from the motion traces of the LH alone. Here, we introduce a *dominant motion rule* for rest determination. We use the motion energy differential of the movements of both hands to determine if small movements in one hand are interpreted as holds. In segment (H), the RH is at rest; hence any movement in the alternate LH becomes significant. In segment (I), the RH exhibits strong motion, and the effects of the LH motion are attenuated.

(J.1.) & (J.2.) *Backstaircase Catchment 1 (Frames 86–132), and Backstaircase Catchment 2 (Frames 753–799):* Figure 8 is a side-by-side comparison of the motions of the right hand that constitute the back staircase description. The subject described the spiral staircase with an upward twirling motion of the RH. In the first case, the subject aborted the gesture and speech sequence with an abrupt interruption and went on to describe the front of the house and the front staircase. In the second case, the subject completes the gestural motion and the back staircase discourse. We can make several observations about this pair of gestures that are separated by more

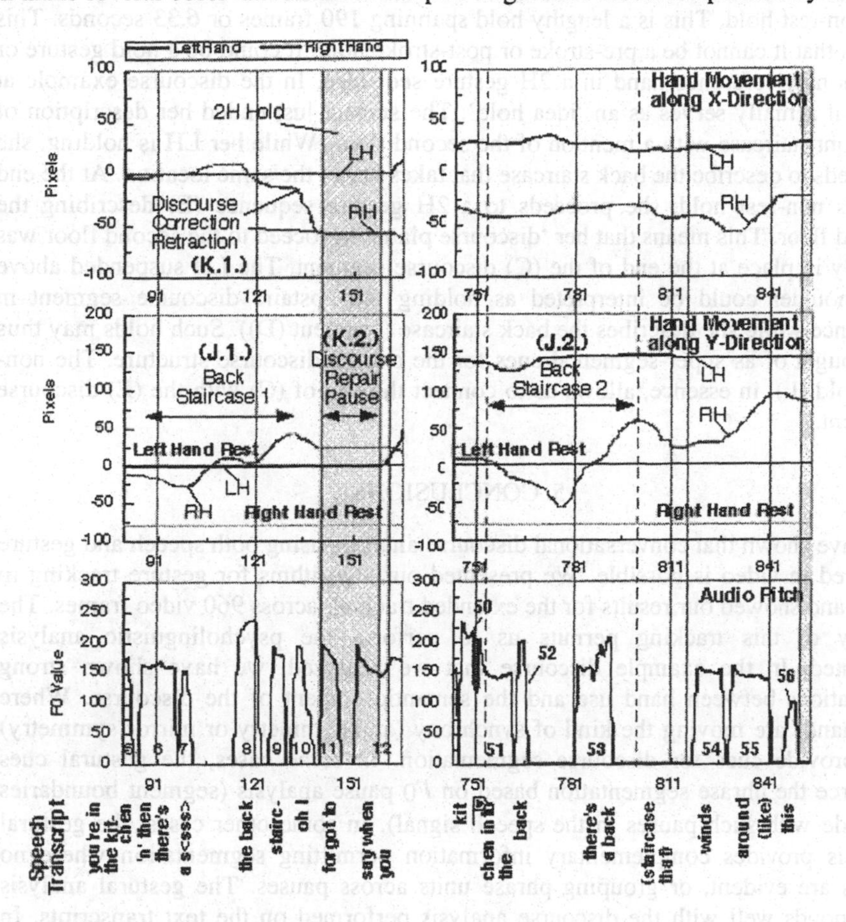

Figure 8. Side-by-side comparison of the back-staircase catchment.

than 22 seconds. First, both are gestures of one hand (RH). Second, the general form of both gestures is similar. Third, up till the discourse repair break, both iteration had exactly the same duration of 47 frames. Fourth, the speaker appears to have already been planning a change in direction in her discourse, and the first motion is muted with respect to the second.

(K.1.) & (K.2.) *Discourse repair retraction (Frames 133-142), and discourse repair pause (Frames 143–159):* Segments (K.1.) and (K.2.) correspond with the speech 'Oh I forgot to say' and flag a repair in the discourse structure. The subject actually pulls her RH back toward herself rapidly and holds an emblematic gesture with an index finger point. While we do not have enough such observations to make a definitive statement, it is likely that abrupt gestural trajectory changes where the hand is retracted from the gestural space suggests a discourse repair.

(L) *Non-rest hold (Frames 740–929):* An interesting phenomenon is seen in the (L) non-rest hold. This is a lengthy hold spanning 190 frames or 6.33 seconds. This means that it cannot be a pre-stroke or post-stroke hold. It could be a hold gesture or a stationary reference hand in a 2H gesture sequence. In the discourse example at hand, it actually serves as an 'idea hold'. The subject just ended her description of the front staircase with a mention of the second floor. While her LH is holding, she proceeds to describe the back staircase that takes her to the same location. At the end of this non-rest hold, she proceeds to a 2H gesture sequence (E) describing the second floor. This means that her 'discourse plan' to proceed to the second floor was already in place at the end of the (C) discourse segment. The LH suspended above her shoulder could be interpreted as holding the upstairs discourse segment in abeyance while she describes the back staircase (segment (D)). Such holds may thus be thought of as super-segmental cues for the overall discourse structure. The non-rest hold (L), in essence, allows us to connect the end of (C) with the (E) discourse segment.

5. CONCLUSIONS

We have shown that conversational discourse analysis using both speech and gesture captured in video is possible. We presented our algorithms for gesture tracking in video and showed our results for the extended tracking across 960 video frames. The quality of this tracking permits us to perform the psycholinguistic analysis presented. In the example discourse that we analyzed, we have shown strong correlations between hand use and the semantic content of the discourse. Where both hands are moving the kind of synchrony (anti-symmetry or mirror symmetry) also provide cues for discourse segmentation. In some cases, the gestural cues reinforce the phrase segmentation based on $F0$ pause analysis (segment boundaries coincide with such pauses in the speech signal). In some other cases, the gestural analysis provides complementary information permitting segmentation where no pauses are evident, or grouping phrase units across pauses. The gestural analysis corresponds well with the discourse analysis performed on the text transcripts. In some cases, the gestural stream provides segmentation cues for linguistic phenomena that are inaccessible from the text transcript alone. The finer grained segmentations in these cases were confirmed by examination of the original videotapes.

We also presented other observations about the gestural signal that are useful for discourse analysis. For some of these observations, we have derived rules for extracting the associated gestural features. We have shown that the final motion preceding a pre-stroke hold should be grouped with the discourse unit of that hold. Likewise, the last movement preceding a rest-hold should be grouped with the rest for that hand. When a hand movement is small, the 'dominant motion rule' permits us to distinguish if it constitutes a hold or gestural hand motion. We have also made observations about repeated motion trajectory catchments, discourse interruptions and repair, and non-hold rests.

Our data also show that while 2D monocular video affords a fair amount of analysis, more could be done with 3D data from multiple cameras. In the two discourse segments labeled (B.1.) and (B.3.) in Figure 6, a significant catchment feature is that the hands move in a vertical plane that is iconic for the surface of the doors. The twirling motion for the back staircase catchment would be more reliably detected with 3D data. The discourse repair retraction toward the subject's body (K.1.) would be evident in 3D. There are also perspective effects that are hard to remove from 2D data. Because of the camera angle (we use an oblique view), a horizontal motion appears to have a vertical trajectory component. This trajectory is dependent on the distance of the hand from the subject's body. In our data, we corrected for this effect by assuming that the hand is moving in a fixed plane in front of the subject. This produces some artifacts that are hard to distinguish without access to the three-dimensional data. In the 'open doors' stroke in (B.2.), for example, hands move in an arc centered around the subject's elbows and shoulders. Hence, there is a large displacement in the 'z' direction (in and away from the body). Hence, the LH seems to move a shorter distance than the right in the x-dimension. The LH also appears to move upward while the RH's motion appears to be almost entirely in the x-dimension. This has directed us to move a good portion of our future experiments to 3D.

We believe that the work reported here just barely scratches the surface of a compelling direction for psycholinguistic research. Such research is necessarily cross-disciplinary, and involves a fair amount of collaboration between the computation sciences and psycholinguistics. The computational requirements for the video computation are significant. On the four-processor (4 X R10000) Silicon Graphics Onyx workstation used for the analyses presented in this paper, we could process 10 seconds of data in 2 hours. We have recently obtained National Science Foundation support for a supercomputer to handle such data and these computationally intensive tasks. Future papers will report the results obtained in 3D utilizing this more powerful computer.

6. ACKNOWLEDGEMENTS

The research and the preparation of this paper has been supported by the U.S. National Science Foundation STIMULATE program, Grant No. IRI-9618887, "Gesture, Speech, and Gaze in Discourse Segmentation", and the National Science Foundation KDI program, Grant No. BCS-9980054, "Cross-Modal Analysis of Signal and Sense: Multimedia Corpora and Tools for Gesture, Speech, and Gaze Research".

7. NOTE

[1] The boldface is the speech with which the stroke coincided. The term 'stroke' is from Kendon, 1980. The stroke is the phase of a gesture with semantic content and the quality of 'effort' in the Labanotation sense. The square brackets indicate the interval when the hand was in motion.

8. REFERENCES

Ansari, R., Y. Dai, J. Lou, D. McNeill, and F. Quek. "Representation of prosodic structure in speech using nonlinear methods", in *1999 Workshop on Nonlinear Signal and Image Processing*, Antalya, Turkey, 1999.

Kendon, Adam. "Gesticulation and speech: Two aspects of the process of utterance", in *The Relationship of Verbal and Nonverbal Communication*, edited by M. Key, 207-27. The Hague: Mouton Publishers, 1980.

Kendon, Adam. "Current issues in the study of gesture", in *The Biological Foundations of Gestures: Motor and Semiotic Aspects*, edited by J-L Nespoulous, P. Peron, and A.R. Lecours, 23–47. Hillsdale, NJ: Lawrence Erlbaum Assoc., 1986.

Kendon, Adam. "How gestures can become like words", in *Crosscultural Perspectives in Nonverbal Communication*, edited by F. Poyatos, 131-41. Toronto: Hogrefe, 1988.

Ladd, D.R., *Intonational Phonology*. Cambridge: Cambridge University Press, 1996.

McNeill, David. *Hand and Mind: What Gestures Reveal about Thought*. Chicago: University of Chicago Press, 1992.

McNeill, David. "Growth points, Catchments, and Contexts." *Cognitive Studies: Bulletin of the Japanese Cognitive Science Society* 7.1 (March 2000): 22-36.

McNeill, David and Susan Duncan. "Growth Points in Thinking-for-Speaking." in *Language and Gesture*, edited by David McNeill, 141-161. Cambridge: Cambridge University Press, 2000.

Nakatani, C.H., B.J. Grosz, D.D. Ahn, and J. Hirschberg, "Instructions for annotating discourses", Technical Report TR-21-95, Ctr for Res. in Comp. Tech., Harvard U., MA, 1995.

Quek, Francis. "Eyes in the Interface." *Int. J. of Image and Vision Comp.* 13 (August 1995): 511-525.

Quek, Francis. "Unencumbered Gestural Interaction." *IEEE Multimedia* (1996) 4: 36-47.

Quek, Francis and R. Bryll, "Vector Coherence Mapping: A parallelizable approach to image flow computation", *Proc. Asian Conf. on Comp. Vis.*, 2 (Jan. 1998): 591–598, Hong Kong.

Quek, Francis, X. Ma, and B. Bryll, "A parallel algorithm for dynamic gesture tracking", *ICCV'99 Wksp on Rec., Anal. & Tracking of Faces & Gestures in R.T. Sys.* (1999): 119–126, Corfu, Greece.

Slobin, D. "Thinking for speaking." In *15th Annual Meeting of the Berkeley Linguistics Society*, pp. 435-445. Berkeley, CA: Berkeley Linguistics Society.

9. AFFILIATIONS

David McNeill and Karl-Erik McCullough: Department of Psychology, University of Chicago, USA.

Francis Quek, Susan Duncan, Robert Bryll, and Xin-Feng Ma: Vision Interfaces and Systems Laboratory, Wright State University, OH, USA.

Rashid Ansari: Signal and Image Research Laboratory, The University of Illinois at Chicago, USA.

DOMINIC W. MASSARO

MULTIMODAL SPEECH PERCEPTION: A PARADIGM FOR SPEECH SCIENCE

1. INTRODUCTION

Speech science evolved as the study of a unimodal phenomenon. Speech was viewed as a solely auditory event, as captured by the seminal speech-chain illustration of Denes & Pinson (1963) shown in Figure 1.

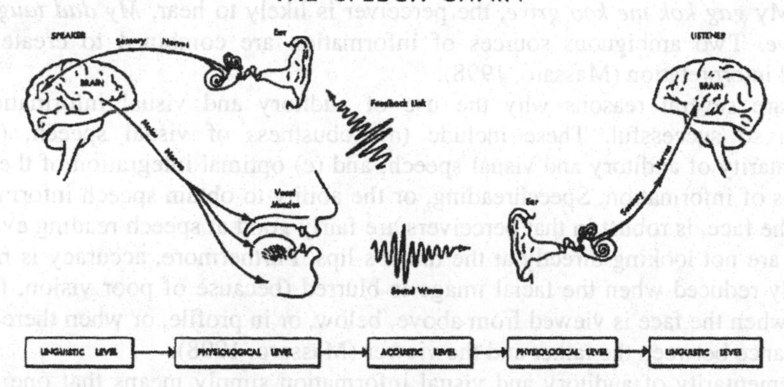

Figure 1. The classic speech-chain illustration of Denes & Pinson (1963).

This view is no longer viable as witnessed by this book as well as a burgeoning record of research findings. Although Denes & Pinson viewed speech as primarily an auditory phenomenon (rather than a multimodal one), they did acknowledge the important contribution of context to accurate recognition and understanding. In accepting the influence of both stimulus information and context on speech perception, the authors anticipated the approach taken in the present chapter. They stated,

> *"In speech communication, then, we do not actually rely on a precise knowledge of specific cues. Instead, we related a great variety of ambiguous cues against the background of the complex system we call our common language."* (Denes & Pinson, 1963, p. 8).

B. Granström et al. (eds.), Multimodality in Language and Speech Systems, 45–71.

Speech as a multimodal phenomenon is supported by experiments indicating that our perception and understanding are influenced by a speaker's face and accompanying gestures, as well as the actual sound of the speech (Massaro, 1987, 1998). Many communication environments involve a noisy auditory channel, which degrades speech perception and recognition. Visible speech from the talker's face (or from a reasonably accurate synthetic talking head) improves intelligibility in these situations. Visible speech also is an important communication channel for individuals with hearing loss.

The number of words understood from a degraded auditory message can often be doubled by pairing the message with visible speech from the talker's face. The combination of auditory and visual speech has been called super-additive because their combination can lead to accuracy that is much greater than accuracy on either modality alone. Furthermore, the strong influence of visible speech is not limited to situations with degraded auditory input. A perceiver's recognition of an auditory-visual syllable reflects the contribution of both sound and sight. For example, if the ambiguous auditory sentence, *My bab pop me poo brive*, is paired with the visible sentence, My *gag kok me koo grive*, the perceiver is likely to hear, *My dad taught me to drive*. Two ambiguous sources of information are combined to create a meaningful interpretation (Massaro, 1998).

There are several reasons why the use of auditory and visual information together is so successful. These include (a) robustness of visual speech, (b) complementarity of auditory and visual speech, and (c) optimal integration of these two sources of information. Speechreading, or the ability to obtain speech information from the face, is robust in that perceivers are fairly good at speech reading even when they are not looking directly at the talker's lips. Furthermore, accuracy is not dramatically reduced when the facial image is blurred (because of poor vision, for example), when the face is viewed from above, below, or in profile, or when there is a large distance between the talker and the viewer (Massaro, 1998).

Complementarity of auditory and visual information simply means that one of the sources is strong when the other is weak. A distinction between two segments robustly conveyed in one modality is relatively ambiguous in the other modality. For example, the place difference between /ba/ and /da/ is easy to see but relatively difficult to hear. On the other hand, the voicing difference between /ba/ and /pa/ is relatively easy to hear but very difficult to discriminate visually. Two complementary sources of information make their combined use much more informative than would be the case if the two sources were non-complementary, or redundant (Massaro, 1998, pp. 424-427).

The final characteristic is that perceivers combine or integrate the auditory and visual sources of information in an optimally efficient manner. There are many possible ways to treat two sources of information: use only the most informative source, average the two sources together, or integrate them in such a fashion in which both sources are used but that the least ambiguous source has the most influence. Perceivers in fact integrate the information available from each modality to perform as efficiently as possible. A wide variety of empirical results has been accurately predicted by a model that describes an optimally efficient process of combination.

In this chapter, I will analyze the multimodality of spoken language understanding within an information-processing framework. After describing the framework, a specific theoretical model is described to help organize the descriptions of experiments and theories. Several alternative theories are then presented and evaluated. To test among the theories, we discuss how the theories account for the influence of multiple sources of stimulus information in speech perception. To structure our information-processing analysis of spoken language understanding, we use a specific theoretical framework that has received substantial support from a variety of experiments in speech perception.

2. THEORETICAL FRAMEWORK

The general theoretical framework provided by the information-processing approach is based on the assumption that there is a sequence of processing stages in spoken language understanding. Stages of information processing have guided, for example, much of the research in visual perception (Palmer, 1999). Visual perception is assumed to occur in three stages of processing: retinal transduction, sensory cues (features), and perceived attributes (DeYoe & Van Essen, 1988). Visual input is transduced by the visual system, a conglomeration of sensory cues is made available, and attributes of the visual world are experienced by the perceiver. In visual perception, there is both a one-to-many and a many-to-one relationship between sensory cues and perceived attributes. The sensory cue of motion provides information about both perceived shape of an object and its perceived movement. A case of the many-to-one relationship in vision is that information about the shape of an object is enriched not only by motion, but also by perspective cues, picture cues, binocular disparity, and shading (e.g., chicariscuro).

We apply this same framework to speech perception and spoken language understanding. Speech perception via the auditory modality is characterized by a transduction of the acoustic signal along the basilar membrane, sensory cues, and perceived attributes. A single sensory cue can influence several perceived attributes. The duration of a vowel provides information about vowel identity (bit vs. beet), information such as lexical stress (the noun and verb pronunciations of the word *permit*), and syntactic boundaries in sentences. Another example is that the pitch of a speaker's voice is informative about both the identity of the speaker and intonation. The best-known example of multiple cues to a single perceived attribute in speech is the case of the many cues for the voicing of a medial stop consonant (Cohen, 1979; Lisker, 1978). These include the duration of the preceding vowel, the onset frequency of the fundamental, the voice onset time, and the silent closure interval. A multimodal example is the impressive demonstration that both the speech sound and the visible mouth movements of the speaker influence perception of place of articulation of a stop consonant (Massaro & Cohen, 1983; McGurk & MacDonald, 1976).

Our research and that of many others has demonstrated a powerful influence of visible speech in face-to-face communication. The influence of several sources of information from several modalities provides a new challenge for theoretical accounts of speech perception. For theories that were developed to account for the perception of unimodal auditory speech (Diehl & Kluender, 1987, 1989), it is not obvious how they would account for the positive contribution of visible speech. Some extant theories view speech perception as a specialized process and not solely as an instance of pattern recognition (Liberman & Mattingly, 1985; Mattingly & Studdert-Kennedy, 1991). We take a different approach by envisioning speech perception as an instance of a more general process of pattern recognition (Massaro, 1998). In language processing, recognition is achieved via a variety of bottom-up and top-down sources of information. Top-down sources include contextual, semantic, syntactic, and phonological constraints; bottom-up sources include audible and visible features of the spoken word. A top-down source might be the overall frequency of a speech segment in the perceiver's language. A bottom-up source might be the degree of jaw rotation while talking.

3. THEORETICAL/EMPIRICAL INQUIRY

Our general framework documents the value of a combined experimental/theoretical approach. The research has contributed to our understanding of the characteristics used in speech perception, how speech is perceived and recognized, and the fundamental psychological processes that occur in speech perception and in pattern recognition in a variety of other domains.

We evaluate the contribution of visible information in face-to-face communication and how it is combined with auditory information in the ecologically valid condition of bimodal speech perception (face-to-face communication). Psychophysical and pattern-recognition tasks are carried out to analyze which audible and visible features are used by human observers in auditory, visual, and auditory-visual (bimodal) speech perception. Quantitative models of feature evaluation and integration are tested against identification judgments, ratings, and confusion matrices from perceptual tests. The results are used to determine which features influence performance.

The results are also to test formal models of speech perception. The models are formalized to make quantitative predictions of the judgments of the test items. Multiple models are tested to preclude a confirmation bias and to adhere to a falsification strategy of inquiry (Massaro, 1989, chapter 5). Each model is tested against the results of single subjects in order to avoid the pitfalls of averaging results across subjects. We also test a variety of participants to explore a broad variety of dimensions of individual variability. These include (1) life-span variability, (2) language variability, (3) sensory impairment, (4) brain trauma, (5) personality, (6) sex differences, and (7) experience and learning. In addition, a large variety of experimental procedures and test situations are used in our investigations (Massaro, 1998, Chapter 6). Generally, we need to know to what extent the processes uncovered in our research generalize across (1) sensory modalities, (2) environmental domains, (3) test items, (4) behavioural measures, (5) instructions, (6) and tasks.

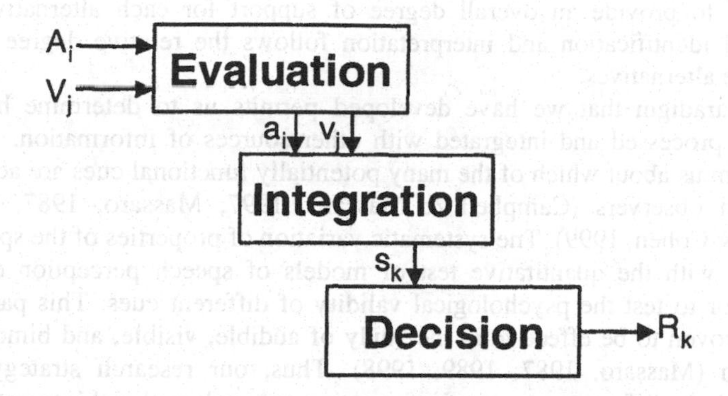

Figure 2. Schematic representation of the three processes involved in perceptual recognition. The three processes are shown to proceed left to right in time to illustrate their necessarily successive but overlapping processing. These processes make use of prototypes stored in long-term memory. The sources of information are represented by uppercase letters. Auditory information is represented by A_i and visual information by V_j. The evaluation process transforms these sources of information into psychological values (indicated by lowercase letters a_i and v_j). These sources are then integrated to give an overall degree of support, s_k, for each speech alternative k. The decision operation maps the outputs of integration into some response alternative, R_k. The response can take the form of a discrete decision or a rating of the degree to which the alternative is likely.

We believe that our empirical work would be inadequate and perhaps invalid without the corresponding theoretical framework. Thus, the research addresses both empirical and theoretical issues. At the empirical level, experiments are carried out to determine how visible speech is combined with auditory speech for a broad range of individuals and across a wide variation of situational domains. At the theoretical level, the assumptions and predictions of several models are formalized, analyzed, contrasted, and tested. Various types of model fitting strategies have been employed, with similar outcomes. These model tests have been highly informative with respect to improving our understanding of how spoken language is perceived and understood.

4. FUZZY LOGICAL MODEL OF PERCEPTION

We have learned that a variety of empirical results can be successfully described within a framework of a fuzzy logical model of perception (FLMP). The FLMP assumes necessarily successive but overlapping stages of processing, as shown in Figure 2. The perceiver of speech is viewed as having multiple sources of information supporting the identification and interpretation of the language input. The model assumes that (1) each source of information is evaluated to give the continuous degree to which that source supports various alternatives, (2) the sources of information are evaluated independently of one another, (3) the sources are integrated to provide an overall degree of support for each alternative, and (4) perceptual identification and interpretation follows the relative degree of support among the alternatives.

The paradigm that we have developed permits us to determine how visible speech is processed and integrated with other sources of information. The results also inform us about which of the many potentially functional cues are actually used by human observers (Campbell & Massaro, 1997; Massaro, 1987, Chapter 1; Massaro & Cohen, 1999). The systematic variation of properties of the speech signal combined with the quantitative test of models of speech perception enables the investigator to test the psychological validity of different cues. This paradigm has already proven to be effective in the study of audible, visible, and bimodal speech perception (Massaro, 1987, 1989, 1998). Thus, our research strategy not only addresses how different sources of information are evaluated and integrated, but can uncover what sources of information are actually used. We believe that the research paradigm confronts both the important psychophysical question of the nature of information and the process question of how the information is transformed and mapped into behaviour. Many independent tests point to the viability of the FLMP as a general description of pattern recognition. The FLMP is centered around a universal law of how people integrate multiple sources of information. This law and its relationship to other laws is developed in detail in Massaro (1998).

The assumptions of the FLMP are testable because they are expressed in quantitative form. The founding or keystone assumption of this model is the division of perception into the twin levels of information and information processing. Adhering to this fundamental dichotomy are a number of other testable assumptions. One is the idea that at the information level, sources are evaluated independently. Independence of sources is motivated by the principle of category-conditional independence (Massaro & Stork, 1998): it is not possible to predict the evaluation of one source on the basis of the evaluation of another, so the independent evaluation of both sources is necessary to make an optimal category judgment. While sources are thus kept separate at evaluation, they are then integrated to achieve perception and interpretation.

Multiplicative integration yields a measure of total support for a given category identification. This operation, implemented in the model, allows the combination of two imperfect sources of information to yield better performance than would be possible using either source by itself. However, the output of integration is an absolute measure of support; it must be relativized, due to the observed factor of relative influence (the influence of one source increases as other sources become less influential, i.e. more ambiguous). Relativization is effected through a decision stage, which divides the support for one category by the summed support for all other categories. An important empirical claim about this algorithm is that while information may vary from one perceptual situation to the next, the manner of combining this information – information processing – is invariant. With our algorithm, we thus propose an invariant law of pattern recognition describing how continuously perceived (fuzzy) information is processed to achieve perception of a category.

5. APPLIED VALUE OF RESEARCH

Many communication environments involve a noisy auditory channel, which degrades speech perception and recognition. Visible speech from the talker's face (or from a reasonably accurate synthetic talking head) improves intelligibility in these situations. Another applied value of visible speech is its potential to supplement other (degraded) sources of information for disabled individuals (Massaro & Cohen, 1999; Oerlemans & Blamey, 1998). Its use is important for hearing-impaired individuals because it allows effective communication within spoken language, the universal language of the community. Just as synthetic auditory speech has been of great importance for research on auditory speech perception, synthetic visual speech is important in studying visual speech perception. In addition, just as auditory speech synthesis has proved a boon to our visually impaired citizens in human machine interaction, visual speech synthesis may prove to be valuable for the hearing impaired. As just one example, cochlear implants have been shown to be successful in allowing implanted individuals to communicate via spoken language. In many situations, however, the electrical speech is not adequate, but the addition of visible speech allows successful communication (Schindler & Merzenich, 1985; Tyler et al., 1992).

It has been estimated by NIDCD that more than twenty-eight million Americans are hearing impaired. It is also the case that roughly three million Americans are estimated to have a corrected visual acuity of 20/40 or worse. With the rapidly increase in the number of elderly people, and the increase in visual and hearing impairment with aging, it is critical that we understand how people process multiple and somewhat ambiguous channels. There is also an unexplored positive potential of visible speech for (1) improving the quality of speech of persons with perception and production deficits, (2) enhancing second language learning and communication, (3) remedial training for poor readers, and (4) human-machine interactions.

6. DEMONSTRATION EXPERIMENT: VARYING THE AMBIGUITY OF THE SPEECH MODALITIES

An important manipulation is to systematically vary the ambiguity of each of the source of information in terms of how much it resembles each syllable. Synthetic speech (or at least a sophisticated modification of natural speech) is necessary to implement this manipulation. In a previous experimental task, we used synthetic speech to cross five levels of audible speech varying between /ba/ and /da/ with five levels of visible speech varying between the same alternatives. We also included the unimodal test stimuli to implement the expanded factorial design, as shown in Figure 3.

6.1. Prototypical Method

The properties of the auditory stimulus were varied to give an auditory continuum between the syllables /ba/ and /da/. In analogous fashion, properties of our animated face were varied to give a continuum between visual /ba/ and /da/. Five levels of audible speech varying between /ba/ and /da/ were crossed with five levels of visible speech varying between the same alternatives. In addition, the audible and visible speech also were presented alone for a total of 25 + 5 + 5 = 35 independent stimulus conditions. Six random sequences were determined by sampling the 35 conditions without replacement giving six different blocks of 35 trials. An experimental session consisted of these six blocks preceded by six practice trials and with a short break between sessions. There were four sessions of testing for a total of 840 test trials (35 x 6 x 4). Thus there were 24 observations at each of the 35 unique experimental conditions. Subjects were instructed to listen and to watch the speaker, and to identify the syllable as /ba/ or /da/. This experimental design was used with 82 participants and their results have served as a database for testing models of pattern recognition (Massaro, 1998).

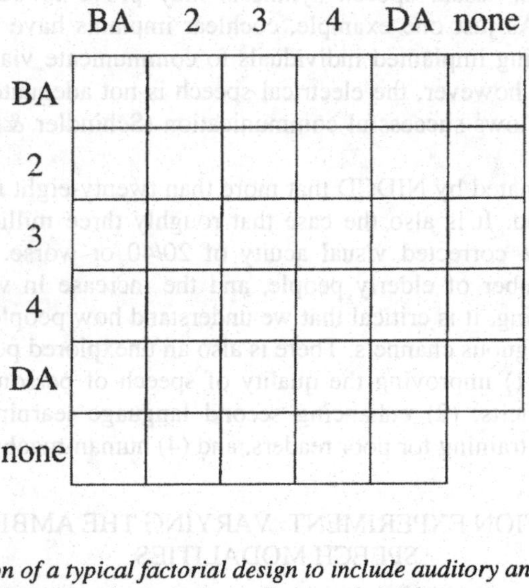

	BA	2	3	4	DA	none
BA						
2						
3						
4						
DA						
none						

Figure 3. Expansion of a typical factorial design to include auditory and visual conditions presented alone. The five levels along the auditory and visible continua represent auditory and visible speech syllables varying in equal physical steps between /ba/ and /da/.

6.2. Prototypical Results

We call these results prototypical because they are highly representative of many different experiments of this type. The mean observed proportion of /da/ identifications was computed for each subject for the 35 unimodal and bimodal conditions. For this tutorial, we present the results for three participants who can be considered typical of the others in this task. The points in Figure 4 give the observed proportion of /da/ responses for the auditory alone (left plot), the bimodal (middle plot), and the visual alone (right plot) conditions as a function of the five levels of the synthetic auditory and visual speech varying between /ba/ and /da/. For the unimodal plots, the degree of influence of a modality is indicated by the steepness of the response function. By this criterion, both the auditory and the visual sources of information had a strong impact on the identification judgments. As illustrated in the left and right plots, the identification judgments changed systematically with changes in the audible and visible sources of information. The likelihood of a /da/ identification increased as the auditory speech changes from /ba/ to /da/, and analogously for the visible speech.

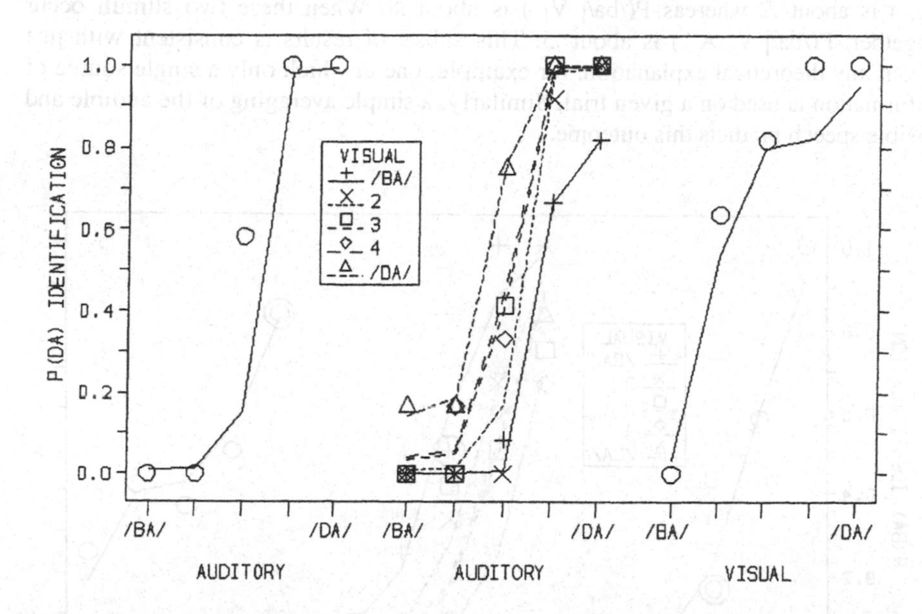

Figure 4. The points give the observed proportion of /da/ identifications for a typical observer in the auditory-alone (left panel), the factorial auditory-visual (center panel) and the visual-alone (right panel) conditions as a function of the five levels of the synthetic auditory and visual speech varying between /ba/ and /da/. The lines give the predictions of the FLMP.

For the bimodal results in the middle plot, the degree of influence is again indexed by the slope of the function for the variable plotted on the x-axis, and by the spread among the curves for the variable described in the key or legend. By these criteria, both sources had a large influence in the bimodal conditions. The curves across changes in the auditory variable are relatively steep and also spread out from on another with changes in the visual variable.

Finally, the auditory and visual effects were *not* additive in the bimodal condition, as demonstrated by a significant auditory-visual interaction. The interaction is indexed by the change in the spread among the curves across changes in the auditory variable. This vertical spread between the curves is about four times greater in the middle than at the end of the auditory continuum. It means that the influence of one source of information is greatest when the other source is neutral or ambiguous. We now address how the two sources are used in perception.

6.3. Evaluation of How Two Sources are Used

Of course, an important question is how the two sources of information are used in perceptual recognition. An analysis of several results informs this question. Figure 5 gives the results for another participant in the task. Three points are circled in the figure to highlight the conditions in which the third level of auditory information is paired with the first (/ba/) level of visual information. When presented alone, P(/ba/ | A_3) is about .2 whereas P(/ba/| V_1) is about .8. When these two stimuli occur together, P(/ba/| V_1 A_3) is about .5. This subset of results is consistent with just about any theoretical explanation, for example, one in which only a single source of information is used on a given trial. Similarly, a simple averaging of the audible and visible speech predicts this outcome.

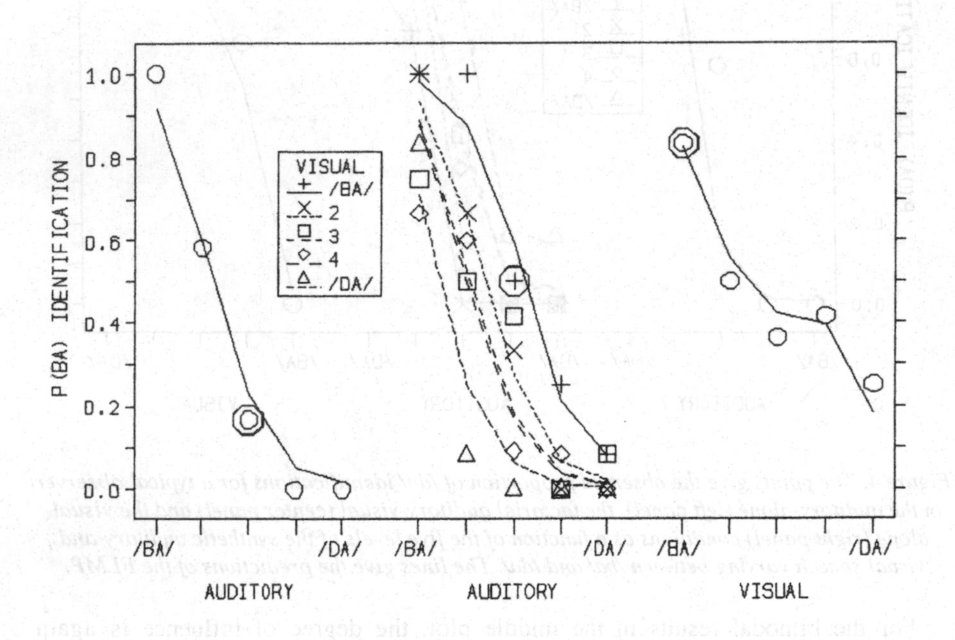

Figure 5. The points give the observed proportion of /ba/ identifications for a typical observer in the auditory-alone (left panel), the factorial auditory-visual (center panel) and the visual-alone (right panel) conditions as a function of the five levels of the synthetic auditory and visual speech varying between /ba/ and /da/. The three circled points A_3 V_1 give two unimodal conditions and the corresponding bimodal condition. The relationship among the three points can be explained by the use of a single modality, an averaging of the two sources, or a multiplicative integration of the two sources. The lines are the predictions of the FLMP.

Other observations, however, allow us to reject these alternatives. Figure 6 gives the results for yet another participant in the task. Three points are circled in the figure to highlight the conditions in which the second level of auditory information is paired with the second level of visual information. When presented alone, P(/ba/ | A_2) is about .8 and P(/ba/| V_2) is about .8. When these two stimuli occur together,

P(/ba/| V_2 A_2) is about 1. This so-called super-additive result (the bimodal is more extreme than either unimodal response proportion) is not easily explained by either the use of a single modality or a simple averaging of the two sources, but is well described by the FLMP. The quantitative predictions of the FLMP have been formalized in a number of different publications (e.g., Massaro, 1987, 1998). In a two-alternative task with /ba/ and /da/ alternatives, the degree of auditory support for /da/ can be represented by a_i, and the support for /ba/ by $(1 - a_i)$. Similarly, the degree of visual support for /da/ can be represented by v_j, and the support for /ba/ by $(1 - v_j)$. The probability of a response to the unimodal stimulus is simply equal to its feature value. For bimodal trials, the predicted probability of a response given auditory and visual inputs, $P(/da/|A_iV_j)$ is equal to

$$P(/da/\mid A_iV_j) = \frac{a_i v_j}{a_i v_j + (1-a_i)(1-v_j)} \quad (1)$$

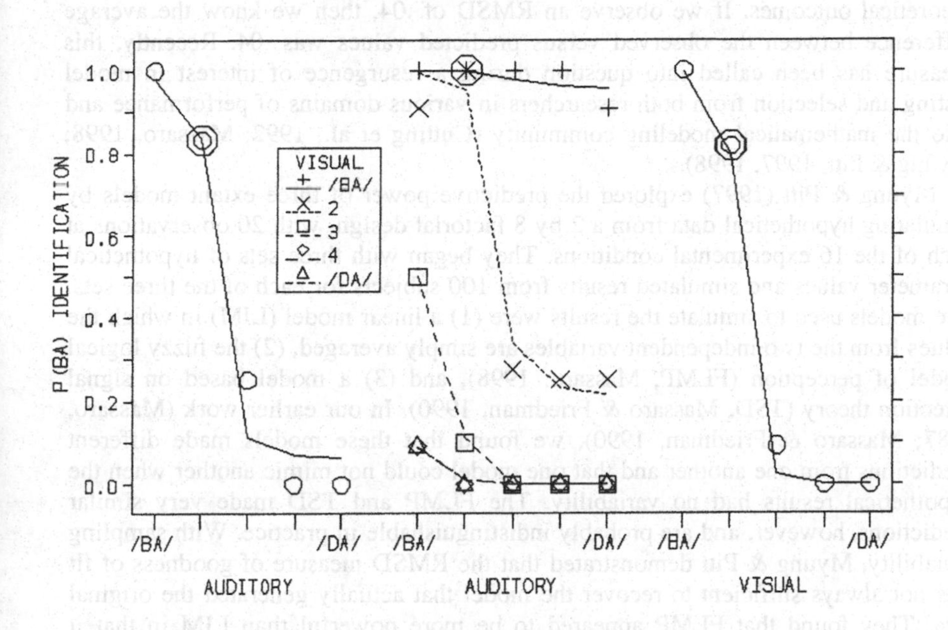

Figure 6. The points give the observed proportion of /ba/ identifications for a typical observer in the auditory-alone (left panel), the factorial auditory-visual (center panel) and the visual-alone(right panel) conditions as a function of the five levels of the synthetic auditory and visual speech varying between /ba/ and /da/. The three circled points A_2V_2 give two unimodal conditions and the corresponding bimodal condition. The relationship among the three points cannot be explained by the use of a single modality or an averaging of the two sources, but can be described by a multiplicative integration of the two sources. The lines are the predictions of the FLMP.

Given that these results using an expanded factorial design and tests of formal models, it is important to replicate this task under a broader set of conditions. These basic findings hold up under a variety of experimental conditions (Massaro, 1998, Chapter 6). In one case, subjects were given just two alternatives, and in the other the same subjects were allowed an open-ended set of alternatives. When tested against the results, the FLMP gives a good description of performance, even with the constraint that the same parameter values are used to describe performance when the number of response alternatives is varied (see Massaro, 1998, pp. 265-268).

7. TESTS OF THE FLMP

We have found that the FLMP has provided the best description of a variety of results in bimodal speech perception. We have contrasted this model against a large number of alternative models. Our criterion for model selection has been the root mean square deviation (RMSD) between predicted and observed values. The RMSD provides an easily understood measure of the agreement between the actual and the theoretical outcomes. If we observe an RMSD of .04, then we know the average difference between the observed versus predicted values was .04. Recently, this measure has been called into question during a resurgence of interest in model testing and selection from both researchers in various domains of performance and also the mathematical modeling community (Cutting et al., 1992; Massaro, 1998; Myung & Pitt, 1997, 1998).

Myung & Pitt (1997) explored the predictive power of three extant models by simulating hypothetical data from a 2 by 8 factorial design, with 20 observations at each of the 16 experimental conditions. They began with three sets of hypothetical parameter values and simulated results from 100 subjects for each of the three sets. The models used to simulate the results were (1) a linear model (LIM) in which the values from the two independent variables are simply averaged, (2) the fuzzy logical model of perception (FLMP, Massaro, 1998), and (3) a model based on signal detection theory (TSD, Massaro & Friedman, 1990). In our earlier work (Massaro, 1987; Massaro & Friedman, 1990), we found that these models made different predictions from one another and that one model could not mimic another when the hypothetical results had no variability. The FLMP and TSD made very similar predictions, however, and are probably indistinguishable in practice. With sampling variability, Myung & Pitt demonstrated that the RMSD measure of goodness of fit was not always sufficient to recover the model that actually generated the original data. They found that FLMP appeared to be more powerful than LIM in that it sometimes gave a better account of the simulated results even when LIM was used to generate the data. When FLMP or TSD was used to generate the hypothetical results, the LIM model never provided a better fit than the other two models. Myung & Pitt (1997) proposed the Bayes factor (Kass & Raftery, 1995) for model selection, which incorporates both functional form and model complexity as criteria for selecting the best model. When applied to the simulated results, this new technique usually provided a recovery of the "correct" model. These results implied that the LIM model might have been erroneously rejected in our previous work (see also Cutting et al., 1992). This is obviously an undesirable state of affairs and challenges our previous conclusions in this arena.

This important analysis and potential solution provided by the Bayes factor alerted us to the possibility that our previous tests between alternative models may have been inadequate. Given the more powerful ability of the FLMP to fit results, even results that were not generated by that model, our conclusions might have been invalid. However, there were several aspects of the Myung & Pitt simulation that did not mirror our prototypical experimental situations. First, the authors simulated data from an unweighted averaging model (LIM) rather than a weighted averaging model (WTAV) that we have tested in all of our research (Massaro, 1998; Massaro & Cohen, 1976). The WTAV is more psychologically realistic in that it is unlikely that each factor is weighted equally in pattern recognition tasks. (This differential weighting in the FLMP emerges from the nonlinear combination of the two sources of information corresponding to the two factors.) Second, the authors simulated data from an asymmetrical factorial design whereas we usually carry out symmetrical expanded factorial designs. The latter are much more powerful than the former in discriminating among different models. A symmetrical design has the best ratio of independent observations to free parameters, and the expanded design provides an additional set of recognition probabilities whose expected values are assumed to be equal to the actual parameter values. Third, the authors used only three hypothetical sets of parameter values whereas we have contrasted the models in literally hundreds of independent tests.

To explore these differences, we carried out a series of comparisons of the use of RMSD versus Bayesian selection in the evaluation of extant models (Massaro et al., 2001). We used a database from the task described in the Demonstration experiment and shown in Figures 4-6 (Massaro et al., 1993; Massaro et al., 1995). This experimental design was used with 82 participants and their results also served as a database for testing models of pattern recognition (Massaro, 1998, Chapters 2 and 10).

For these 82 participants, the FLMP gave a better description than the WTAV model for 94% of the real subjects. To analyze the robustness of the RMSD measure, we created a set of hypothetical subjects who behaved according to either one model or the other. These montecarlo simulations involved creating 20 simulated subjects for each model for each real subject. By using the same number of trials, the simulation should have the same sampling variability as was present in the data set being modelled. For these simulated participants, the RMSD measure was sufficient to recover the original model that generated the data. For both data sets, the incorrect model was recovered only 1% of the time. These same results were used to test the models on the basis of the Bayes factor; Kass & Raftery, 1995) for model selection, which incorporates both functional form and model complexity as criteria for selecting the best model. When applied to these empirical results, this new technique did not change the conclusions that were reached. The FLMP maintained its significant descriptive advantage over the WTAV with this new criterion (Massaro et al., 2001). The outcomes support the conclusion that the RMSD measure yields similar outcomes to the Bayes factor for the conditions of our prototypical design. Thus, the validity of the FLMP holds up under even more demanding methods of model selection.

As in all things, there is no holy grail of model evaluation for scientific inquiry. As elegantly concluded by Myung & Pitt (1997), the use of judgment is central to model selection. Extending their advice, we propose that investigators should make use of as many techniques as feasible to provide converging evidence for the selection of one model over another. More specifically, both RMSD and the Bayes factor can be used as independent metrics of model selection. Inconsistent outcomes should provide a strong caveat for the validity of selecting one model over another in the same way that conflicting sources of information create an ambiguous speech event for the perceiver.

8. CLARIFIYING THE MCGURK EFFECT

It has been well over two decades since the publication of the McGurk effect (McGurk & MacDonald, 1976), which has obtained widespread attention in many circles of psychological inquiry and cognitive science (Green, 1998; Schwartz et al., 1998). The classic McGurk effect involves the situation in which an auditory /ba/ is paired with a visible /ga/ and the perceiver reports hearing /da/, called a fusion response. The reverse pairing, an auditory /ga/ and visual /ba/, tends to produce a perceptual judgment of /bga/, called a combination response. The finding that auditory experience is influenced by the visual input stimulated many students of speech perception to carry out similar investigations. However, many previous studies used just a few experimental conditions in which the auditory and visual sources of information are made to mismatch. Many experiments failed to test the unimodal conditions separately so that there is no independent index of the perception of the single modalities. The experiments also tend to take too few observations under each of the stimulus conditions. The data analysis is also usually compromised because investigators analyze the data with respect to whether or not there was a McGurk effect, which often is simply taken to mean whether the visual information dominated the judgments. This analysis can be highly misleading because we have seen in Figures 4-6 that one modality does not dominate the other. Both modalities contribute to the perceptual judgment with the outcome that the least ambiguous source of information has the most influence. I propose a better understanding of the McGurk effect by enhancing the database and testing formal models of the perceptual process.

To explore the McGurk effect more fully, we carried out a series of experiments in which the auditory syllables /ba/, /da/, and /ga/ were crossed with these same visible syllables in an expanded factorial design. Subjects are either limited to these three response alternatives or given a larger set of response alternatives. Why does auditory /ba/ paired with a visible /ga/ produce a perceptual report of hearing /da/ rather than /ga/? Initial explanation of this outcome has been to expect it to follow from the psychophysical properties of the audible and visible sources of information. This means that visual da/ and visual /ga/ are virtually indistinguishable and that auditory /ba/ must be somewhat more similar to an auditory /da/ than to an auditory /ga/.

Another possibility is that there are other sources of information (or constraints) contributing to the preference of /da/ over /ga/. One of the themes of our research is that there are multiple influences (both top-down and bottom-up) on perceptual processing. One potential top-down source is the frequency of occurrence of these segments in the language. Previous studies have shown that transitional probability contributes to perceptual processing (Massaro & Cohen, 1983; Pitt & McQueen, 1998), and word frequency has been shown to be highly functional in word recognition. Top-down context might be functional in the McGurk effect because the segment /d/ appears to be more frequent in initial position than the segment /g/ (Denes, 1963). This a priori bias for /d/ over /g/ (and /t/ over /k/) could be an important influence contributing to the "fusion" response that is observed.

To explore these two contributions, the natural auditory syllables /ba/, /da/, and /ga/ were crossed with the synthetic visual syllables /ba/, /da/, and /ga/. Participants also identified the unimodal syllables. Ten participants were tested for two sessions of 216 trials each, for a total of roughly 29 observations under each of the 15 conditions. Subjects were given the response alternatives /ba/, /da/, /ga/ or were permitted to make combination responses involving these alternatives (e.g., /bga/).

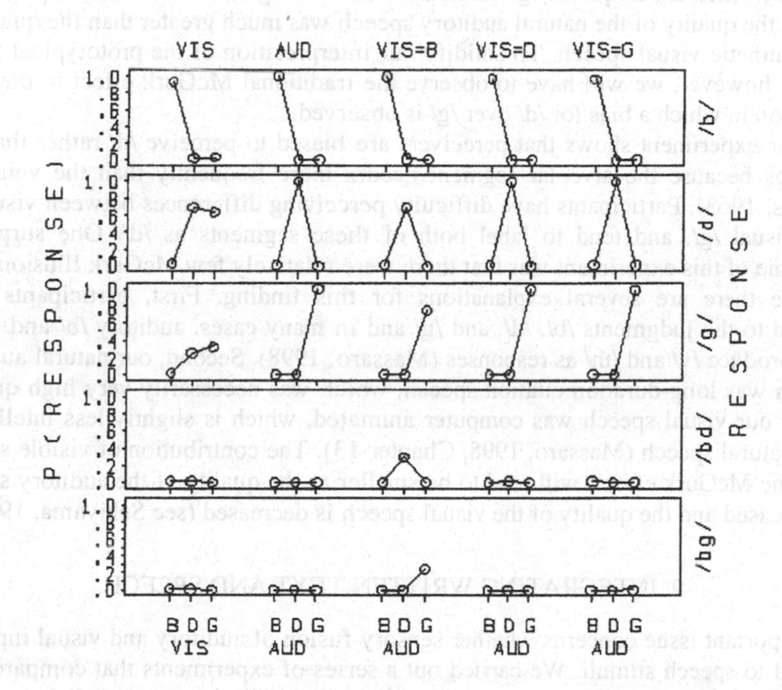

Figure 7. The percentage of /b/, /d/, /g/, /bd/, and /bg/ responses as a function of the three test stimuli in the unimodal visual (VIS), unimodal auditory (AUD), and bimodal conditions.

Figure 7 gives the probability of /ba/, /da/, /ga/, /bda/, and /bga/ responses for each of the 15 experimental conditions. Several results are of interest. As expected, there were confusions between visible /da/ and /ga/, because these syllables tend to look the same. Their major differences in articulation occur inside the mouth, which are hidden to the perceiver. As can be seen in the left plot of Figure 6, the visual syllable /da/ was identified as /d/ about as often as it was identified as /g/. The same result occurred for the syllable /ga/. What is important for our purposes, however, is that the participants respond to both visual /da/ and visual /ga/ about twice as often with the alternative /da/ than with the alternative /ga/. This observation is novel because previous investigators had not tested these unimodal visual conditions. This result offers a new explanation of why an auditory /ba/ paired with a visual /ga/ produces the response /da/. Apparently, people are biased to report /d/ over /g/ because /d/ occurs much more often than /g/ in spoken language (Denes, 1963).

Much to our dismay, however, we failed to find a strong McGurk fusion effect. Neither a visual /da/ or /ga/ biased the response to auditory /ba/. For whatever reason, the auditory information tended to dominate the perceptual judgment. One possibility is that observers were not permitted to make other responses, such as /va/ or /tha/, which are frequently given to these conflicting syllables. Another possibility is that the quality of the natural auditory speech was much greater than the quality of the synthetic visual speech. To solidify our interpretation of the prototypical fusion effect, however, we will have to observe the traditional McGurk effect in the same situation in which a bias for /d/ over /g/ is observed.

Our experiment shows that perceivers are biased to perceive /d/ rather than /g/, perhaps because the alveolar segment occurs more frequently than the velar one (Denes, 1963). Participants have difficulty perceiving differences between visual /d/ and visual /g/, and tend to label both of these segments as /d/. One surprising outcome of this experiment was that there were relatively few McGurk Illusions. We believe there are several explanations for this finding. First, participants were limited to the judgments /b/, /d/, and /g/ and in many cases, auditory /b/ and visual /d,g/ produce /v/ and /th/ as responses (Massaro, 1998). Second, our natural auditory speech was long-duration citation speech, which was necessarily very high quality. Third, our visual speech was computer animated, which is slightly less intelligible than natural speech (Massaro, 1998, Chapter 13). The contribution of visible speech (i.e., the McGurk effect) will tend to be smaller as the quality of the auditory speech is increased and the quality of the visual speech is decreased (see Sekiyama, 1998).

9. INTEGRATING WRITTEN TEXT AND SPEECH

An important issue concerns whether sensory fusion of auditory and visual inputs is limited to speech stimuli. We carried out a series of experiments that compared the perception of auditory speech paired with visible speech versus auditory speech paired with written language. The results from this study can help inform us about which theories of bimodal speech perception are viable. Knowing or seeing the words to a rock song while hearing the song creates the impression of hearing a highly intelligible rendition of the words. Without this knowledge of the words, the listener cannot make heads or tails of the message. The first demonstration of this kind that we know of was by John Morton, who played a song by the Beatles. Members of the audience could not perceive clearly the words of the song until they were written on the viewing screen. Another variation on this type of illusion is the so-called phonemic restoration effect in which we claim to hear the /s/ in the word legislatures even though it is replaced by a cough, a buzz or even a pure tone (Warren, 1970).

Frost et al. (1988) found that when a spoken word is masked by noise having the same amplitude envelope, subjects report that they hear the word much more clearly when they see the word in print at the same time. This result supports the idea that written text can influence our auditory experience. To show effects of written information on auditory judgment at the perceptual level, Massaro et al. (1988) compared the contribution of lip-read information to written information. Subjects were instructed to watch a monitor and listen to speech sounds. The sounds were randomly selected from nine synthetic speech sounds along a /ba/ to /da/ continuum. On each trial, the subjects were presented with either (1) a visual representation of a man articulating the sound /ba/ or /da/, or (2) a written segment BA or DA. Although there was a large effect of visible speech, there was only a small (but significant) effect of the written segments on the judgments. Both the speech and written-text conditions were better described by the FLMP than by an alternative additive or single channel model.

To better test for the possible influence of text on speech perception, our study tested whether we could obtain a larger effect of written text. Given that letters of the alphabet have a strict spelling-to-sound mapping and are pronounced automatically and effortlessly, the letters B and D were used. The letter sequences BA and DA are not necessarily pronounced /ba/ and /da/. The letters B and D are only pronounced /bi/ and /di/ – as they are named in the alphabet.

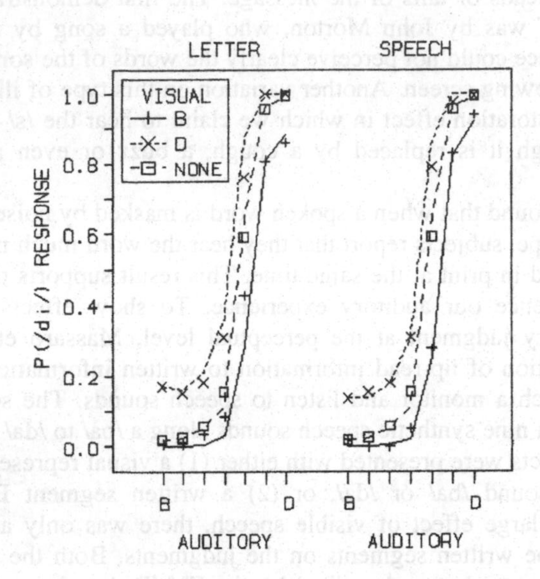

Figure 8. Observed (points) and predicted (lines) by the FLMP probability of a /di/ response as a function of the auditory and visual stimuli for the letter and word conditions.

Nine participants from the University of California, Santa Cruz, were tested. This experiment employed a within-subjects expanded factorial design. There were seven auditory levels between the syllables /bi/ and /di/. There were four visual levels – two letter conditions (the letters B and D) and two speech conditions (the visual syllables /bi/ and /di/), for a total of 39 trial types. The observers were specifically instructed to both watch the screen and listen for a sound and to report what they heard. On those trials in which only a visual stimulus was presented, they were to report the visual stimulus. On each trial, subjects identified stimuli as B or D by typing the appropriately marked keys. The stimuli were presented in 6 blocks of the 39 trial types, for a total of 234 trials per session. The test conditions were selected at random without replacement. A practice block of 10 trials occurred prior to the experimental trials. Subjects had approximately three seconds to respond on each trial. Each subject participated on two days with two sessions per day. Thus there were 24 observations per subject per condition. The dependent measure was the proportion of /di/ judgments.

Figure 8 displays the average results for the letter and speech conditions. The proportion of /di/ responses as a function of the seven auditory levels is shown with the visual B or D stimulus or no visual information (NONE) as the curve parameter. The average proportion of /di/ responses increased significantly as the auditory syllable went from the most /bi/-like to the most /di/-like level. There was also a significant effect on the proportion of /di/ responses as a function of the visual stimulus; there were fewer /di/ responses for visual B than for a visual D. The interaction of these two variables was also significant: the influence of the visual variable was larger at the more ambiguous regions of the auditory continuum. Not shown in Figure 8, the visual alone trials gave essentially perfect performance for both the speech and letters.

The result of interest here is the difference between the visible speech and the letter conditions. As can be seen in the figure, the visual effect was substantial and of similar size for the letter and for the speech condition. The FLMP was fit to the average proportion of /di/ responses for each of the nine participants. The FLMP gave a very good description of the observations. Thus, it appears that written text, as well as visible speech, can influence our auditory experience and that the FLMP accounts for both types of influence. Given these results, it is important to explore a number of important variables to test whether there are any qualitative differences between the integration of written text or visual speech with auditory speech. These conclusions hold up in additional studies with a larger number of response alternatives and without visual-alone trials. Unless we are completely wedded to the idea that speech is special, an influence of written language on our perceptual experience should not be surprising.

10. WORD RECOGNITION

We believe that visual input is a strong influence on spoken language perception in face-to-face communication. An important issue is a possible concern that research with syllables might not generalize to words and sentences. Experimental results with syllables should be compared with those with words and sentences to determine if the same model can be applied to these different test items. To move beyond syllables, we assessed the processing of auditory and visual speech at the word level. Settling on an experimental task for evaluation is always a difficult matter. Even with adequate justification, however, it is important to see how robust the conclusions are across different tasks. In one experiment, we used a gating task, in which successively longer portions of a test word are presented (Grosjean, 1980; Munhall & Tohkura, 1998).

Following our theoretical framework, we tested observers under auditory, visual, and bimodal conditions. The test words were monosyllabic CVCs. Eight gating durations were tested. We expected performance to improve with increases in the duration of both the auditory and visual components of the test word. We expect the auditory information to be more informative than visual, but most importantly bimodal performance should be significantly better than either unimodal condition.

The results were as expected. The FLMP and competing models were fit to both the accuracy of identification of the test words, as well as to the identification of the individual segments of the word. The FLMP gave the best description of the results. This extension of our paradigm to words is an important test of how well our theoretical framework applies beyond the syllable level.

11. PERCEPTION OF PARALINGUISTIC INFORMATION

Laypersons and researchers agree that communication involves much more than simply the linguistic message. Paralinguistic as well as linguistic information is necessary for optimal communication and understanding. Our research has been directed primarily at the linguistic dimensions of speech but it is important that we explore the paralinguistic ones in parallel. In collaboration with Jonas Beskow from KTH, we studied the joint influence of F0, loudness, eye widening and eyebrow movements on the perception of stress. A stressed word tends to be somewhat longer in duration, somewhat greater amplitude, and somewhat higher in pitch. Cave et al. (1996) found that rapid rising-falling eyebrow movements occurred with F0 rises about 70% of the time. There is some other unpublished evidence that eye widening might occur on stress words. Using our factorial design methodology, we manipulate these sources of information independently of one another to determine their relative contributions to the perception of stress. Participants were asked to indicate the degree to which a given word in a sentence was stressed. The analyses included tests of formal models of speech perception and language processing (see Massaro, 1996, 1998).

In one experiment carried out in collaboration with Jonas Beskow, participants were given sentences of the form noun-verb-noun, and asked to indicate whether the first or last word was emphasized. Four independent variables were orthogonally varied in a factorial design. Using an animated talking head and synthetic speech, the eyebrows were raised during either the first or last word, the eyes were widened during either the first or last word, the amplitude was increased during either the first or last word, and the pitch was raised during the first or last word or was held constant during both of the words. For the all four variables, the noun that did not receive emphasis for a given variable was set at the neutral value for that variable. For example, if the eyebrows are raised during the last noun, the eyebrows are kept still during the first noun of that sentence. For the F0 variable, there was a third condition in which the pitch could be kept neutral during both nouns. This gives a 2 by 2 by 2 by 3 factorial design for a total of 24 experimental conditions. The experiment consisted of 20 noun-verb-noun sentences, each presented under each of the 24 experimental conditions, yielding a total of 480 trials. Stimuli were presented in random order, with a short break half way through the experiment. Nine subjects were tested in the experiment, which took place in front of a computer screen, where stimuli were presented by the animated face. Subjects entered their responses using the mouse by clicking on the noun that they perceived to be more stressed in a text representation of the sentence that was presented below the face.

Figure 9 presents the results of the experiment in terms of the proportion of times the first noun in the sentence was categorized as stressed. As can be seen in the figure, although all four independent variables had some influence on the judgments, the amplitude of the noun was the most influential factor.

This situation is slightly more complicated than the prototypical experiment and it will be worthwhile to describe how the FLMP and alternative models can be applied to the results. It is assumed that perceivers evaluate and integrate a variety of cues to perceive word stress. As a working hypothesis, it is assumed that the perceivers evaluate the four cues eyebrow raising (ER), eye widening (EW), amplitude increase (AI), and F0 raising (F0) as cues to stress. A stressed word S(word), can be represented by

$$S(\text{word}): ER \,\&\, EW \,\&\, AI \,\&\, F0 \,\&\, O$$

where O corresponds to other potential cues that are not being systematically manipulated in the experiment. An unstressed word, ~S(word), would be represented by

$$\sim S(\text{word}): \sim ER \,\&\, \sim EW \,\&\, \sim AI \,\&\, \sim F0 \,\&\, \sim O$$

It is assumed that the perceiver determines the degree of stress and unstress for both the first and last nouns in the sentences. The probability of choosing the first noun as stressed is determined by the degree to which the first noun is stressed and the degree to which the second noun is unstressed. It is possible to eliminate the O term corresponding to the other cues since they do not change under the different conditions.

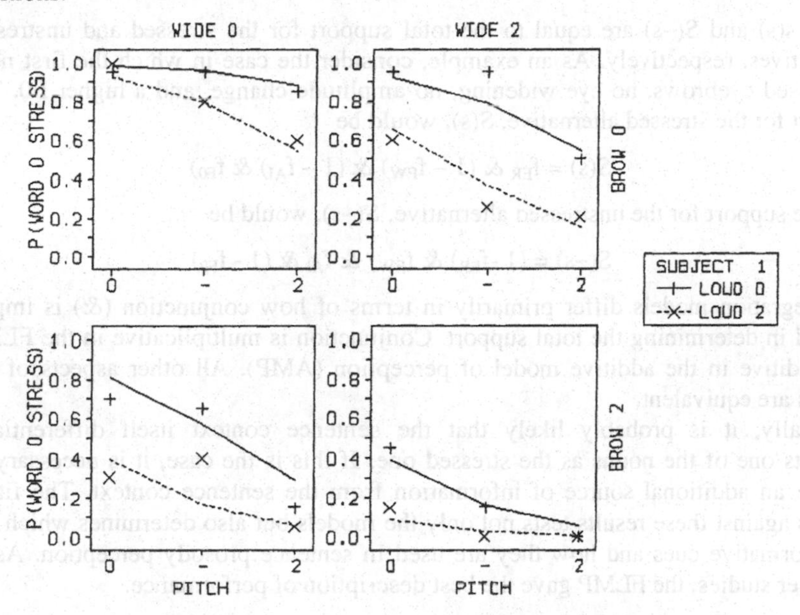

Figure 9. Proportion of times the first word was categorized as stressed, P(Word 0 Stress), as a function of whether the eyebrows were raised during either the first (0) or last word (2), the eyes were widened during either the first or last word, the amplitude was increased during either the first or last word, and the pitch was raised during the first or last word or was held constant (-) during both of the words.

Because the properties of the second noun are simply the unmarked complements of those in the first noun, it is sufficient to simply determine the amount of support for the first noun being stressed or unstressed. The degree to which the first noun is stressed is therefore directly related to having the four cues marking stress in the first noun. If a stimulus cues matches its representation in the prototype for stress, we give it a feature value of f_i; if it mismatches the representation for stress, we give it a feature value of $(1 - f_i)$. The index i stands for the four cues manipulated in the experiment. The value f_i should be greater than .5 if the stress marking is a functional cue to stress and its value can be interpreted as its cue strength. The F0 neutral condition in which both nouns have neutral stress should be given the value .5, which represents the complete absence of support in one direction or the other. Imposing this constraint also insures that the estimated parameter values given the model fit are identifiable or unique (see Crowther et al., 1995; Massaro, 1998, Chapter 11).

The probability of a stressed judgment is equal to the support for a stressed alternative divided by the sum of support for stress and for unstressed alternatives.

$$P(S) = \frac{S(s)}{S(s) + S(\sim s)} \qquad (2)$$

where s(s) and S(~s) are equal to the total support for the stressed and unstressed alternatives, respectively. As an example, consider the case in which the first noun has raised eyebrows, no eye widening, no amplitude change, and a higher F0. The support for the stressed alternative, S(s), would be

$$S(s) = f_{ER} \; \& \; (1 - f_{EW}) \; \& \; (1 - f_{AI}) \; \& \; f_{F0})$$

The support for the unstressed alternative, S(~s), would be

$$S(\sim s) = (1 - f_{ER}) \; \& \; f_{EW}) \; \& \; f_{AI} \; \& \; (1 - f_{F0})$$

Integration models differ primarily in terms of how conjunction (&) is implemented in determining the total support. Conjunction is multiplicative in the FLMP and additive in the additive model of perception (AMP). All other aspects of the models are equivalent.

Finally, it is probably likely that the sentence context itself differentially supports one of the nouns as the stressed one. If this is the case, it is necessary to assume an additional source of information from the sentence context. The fit of models against these results tests not only the models but also determines which are the informative cues and how they are used in sentence prosody perception. As in our other studies, the FLMP gave the best description of performance.

12. PERCEIVING EMOTION FROM THE FACE AND THE VOICE

We have also carried out a set of studies of how people evaluate and integrate information in the recognition of emotion. Our research has varied multiple sources of paralinguistic information, facial expressions and vocal cues, to investigate perception of a speaker's emotion. We have undertaken a set of studies to assess what properties of the face and voice people actually use to infer emotional content. As for speech, we operate under the assumption that multiple sources of information are also used to perceive a talker's emotion. A variety of signals are used in addition to the verbal content of the speech. The emotion may be interpreted in different ways depending on the voice quality, facial expression, and body language used. In order to study the degree to which emotional sources of information are used, it is important to define these sources and then determine how they are used. In our research, two sources of emotional information, facial expressions and vocal cues, are chosen to be as analogous to the speech situation as possible.

In previous research (Ellison & Massaro, 1996), we used an expanded-factorial design where the affective categories happy and angry were chosen because they represent two of the basic categories of emotion. We chose two features that seem to differ somewhat in happy and angry faces. An important criterion for manipulating two features is that they can be varied independently of one another. Thus, varying one cue in the upper face and one cue in the lower face was an ideal solution. Five levels of the upper face and five levels of the lower face were factorially combined, along with the ten half-face conditions. The feature values were obtained by comparison to features displayed in exemplar photographs in Ekman & Friesen (1975). The features varied were brow deflection (BD) and mouth-corner deflection (MD). BD was varied from fully elevated and arched for a prototypically happy affect to fully depressed and flattened for a prototypically angry affect. MD was varied from fully curled up at corners for a prototypically happy affect to fully curled down at corners for a prototypically angry affect. The FLMP gave a good fit to individual fits in binary judgments, judgments with six response alternatives, and in a rating task in which with instructions to rate the affect on a scale from 1 to 9. The independent variables influenced performance in the same manner in all tasks.

In another experiment, we examined how emotion is perceived by using facial and vocal cues of a speaker (Massaro & Egan, 1996). Three levels of facial affect were presented using a computer-generated face. Three levels of vocal affect were obtained by recording the voice of a male amateur actor who spoke a semantically neutral word in different simulated emotional states. These two independent variables were presented to participants of the experiment in all possible permutations, i.e. visual cues alone, vocal cues alone and visual and vocal cues together, which gave a total set of 15 stimuli. The participants were asked to judge the emotion of the stimuli in a two-alternative forced choice task (either HAPPY or ANGRY). The results indicate that participants evaluate and integrate information from both modalities to perceive emotion. The influence of one modality was greater to the extent that the other was ambiguous (neutral). The FLMP fit the judgments significantly better than an additive model, which weakens theories based on an additive combination of modalities, categorical perception, and influence from only a single modality.

We have extended these studies to the other basic emotions, including fear, surprise, disgust, and sadness. These studies not only tested how well the theoretical framework holds up with these other emotions, they also provide an understanding of which facial features are informative in conveying these emotions. As an example, there are cues in the upper and lower face that can be varied to create surprised and fearful emotions. A surprised face is characterized by a wide-eyed look with very raised eyebrows and an open mouth. A fearful face, on the other hand, has a somewhat less raising of the eyebrows and less opening of the mouth. Given that the FLMP provides a good description of performance for the full set of basic emotions, the perception of emotion appears to be well described by our theoretical framework.

13. TESTS OF DYNAMIC INFORMATION IN VISIBLE SPEECH PERCEPTION

It has been proposed that speech perception is based on higher order dynamic properties of the spoken language. Point-light displays have been used as support for this idea. Rosenblum & Saldana (1996, 1998) using hybrid visual-auditory tests have argued that a point-light display using 28 lights attached to the face is effective as a visual speech stimulus. The authors claim that point-light displays were effective for the /b/-/v/ distinction. However, they did not convincingly demonstrate that it was as informative as a real moving face. We have tested this idea by comparing point-light displays to the perception of displays of our synthetic talker (Cohen, et al., 1996). This test provides a more extensive assessment of how much information can be transmitted by point-light displays, as well as a better control to equate the stimuli for the two conditions. Rosenblum & Saldana's original study (1) used only /ba/ and /va/ tokens, and (2) the facial and point-light displays were made separately under differing conditions, which makes it difficult to say whether the stimuli had equivalent articulations.

The experimental procedure was identical to that used in Massaro (1998, chapter 13), except that the synthetic point-light face replaced the natural one. The point-light display was made by putting tiny spheres on 28 of the polygon vertices of the face, positioned on the face, lips, teeth, and tongue identically to those used by Rosenblum & Saldana. Performance on the initial consonant visemes of the test words was significantly worse for the point-light display compared to our synthetic facial display. Analysis of performance on final consonant and vowel visemes also showed a significant advantage for the synthetic face over the point-light display. It is evident that with equivalent display geometries, the point-light display does provide some valuable information for speech reading, although performance was significantly worse than the full synthetic facial display. Although it might be the case that kinematic properties are informative for speech reading, our results indicate that they are not sufficient.

14. CONCLUSION

If the reader has persisted to this stage, it is hoped that they have obtained a good understanding of our approach to the study of speech perception and understanding. Like multimodality, which provides several roads to understanding speech, our framework exploits multiple ways of knowing about how this magnificent and magical process works. One goal was to illustrate the value of an information-processing framework for the study of multimodality in spoken language understanding. Our theoretical framework and the FLMP were used to impose coherence on a complex set of experiments and results. I look forward to the future in which the concept of multimodality is being applied fruitfully in theoretical development, empirical research, applied situations, and commercial endeavors.

15. REFERENCES

Campbell, C.S. & D.W. Massaro. "Perception of visible speech: influence of spatial quantization", *Perception, 26*, 627-644, 1997.

Cave, C., I. Guaitella, R. Bertrand, S. Santi, F. Harlay & R. Espesser. "About the relationship between eyebrow movements and FO variations". *Proceedings of the International Conference on Spoken Language Processing* (pp. 2175-2178), Wilmington: University of Delaware, 1996.

Cohen, M.M., R.L. Walker & D.W. Massaro. "Perception of synthetic visual speech". In: D.G. Stork & M.E. Hennecke (Eds.), *Speechreading by humans and machines* (pp. 153-168). New York: Springer, 1996.

Cole, R., T. Carmell, P. Connors, M. Macon, J. Wouters, J. deVilliers, A. Tarachow, D.W. Massaro, M.M. Cohen, J. Beskow, J. Yang, U. Meier, A. Waibel, P. Stone, G. Fortier, A. Davis, C. Soland. "Intelligent Animated Agents for Interactive Language Training". *Proceedings of Speech Technology in Language Learning*. Stockholm, Sweden, 1998.

Crowther, C.S., W.H. Batchelder & X. Hu. "A measurement-theoretical analysis of the Fuzzy Logical Model of Perception". *Psychological Review, 102*, 396-408, 1995.

Cutting, J.E., N. Bruno, N.P. Brady & C. Moore. "Selectivity, scope, and simplicity of models: A lesson from fitting judgments of perceived depth". *Journal of Experimental Psychology: General*, 121, 364-381, 1992.

Denes, P.B. "On the statistics of spoken English". *Journal of the Acoustical Society of America, 35*, 892-904, 1963.

Diehl, R.L. & K.R. Kluender. "On the categorization of speech sounds". In: S. Harnad (Ed.), *Categorical perception* (pp. 226-253). Cambridge: Cambridge University Press, 1987.

Diehl, R.L. & K.R. Kluender. "On the objects of speech perception". *Ecological Psychology*, 121-144, 1989.

DeYoe, E.A. & D.C. Van Essen. "Concurrent processing streams in monkey visual cortex". *Trends in Neurosciences, 11*, 219-226, 1988.

Ekman, P. & W. Friesen. *Pictures of facial affect*. Palo Alto, CA: Consulting Psychologists Press, 1975.

Ellison, J.W. & D.W. Massaro. "Featural evaluation, integration, and judgement of facial affect", *Journal of Experimental Psychology: Human Perception and Performance, 2*, 213-226, 1997.

Fowler, C.A. "Listeners do hear sounds, not tongu". *Journal of the Acoustical Society of America, 99*, 1730-1741, 1996.

Frost, R., B.H. Repp & L. Katz. "Can speech perception be influenced by simultaneous presentation of print?" *Journal of Memory and Language, 27*, 741-755, 1988.

Green, K.P. "The use of auditory and visual information during phonetic processing: Implications for theories of speech perception". In: Campbell, R., B. Dodd & D. Burnham (Eds.), *Hearing by Eye II* (pp. 3-25). East Sussex, UK: Psychology Press Ltd, 1998.

Grosjean, F. "Spoken word recognition processes and the gating paradigm". *Perception & Psychophysics, 28*, 267-283, 1980.

Kass, R.E. & A.E. Raferty. "Bayes factors". *Journal of the American Statistical Association, 90*, 773-795, 1995.

Liberman, A.M. & I.G. Mattingly. "The motor theory of speech perception revised". *Cognition, 21*, 1-33, 1985.

Lisker, L. "Rabid vs rapid: A catalog of acoustic features that may cue the distinction". *Haskins Laboratories, Status Report on Speech Research, SR-54*, 127-132, 1978.

Massaro, D.W. *Speech Perception by Ear and Eye: A Paradigm for Psychological Inquiry*. Hillsdale, NJ: Lawrence Erlbaum Associates, 1987.

Massaro, D.W. Multiple book review of *Speech perception by ear and eye: a paradigm for psychological inquiry*, by D.W. Massaro. *Behavioral and Brain Sciences, 12*, 741-794, 1989.

Massaro, D.W. "Integration of multiple sources of information in language processing". In: T Inui & J.L. McClelland (Eds.), *Attention and Performance XVI: Information integration in perception and communication* (pp. 397-432). Cambridge, MA: MIT Press, 1996.

Massaro, D.W. *Perceiving Talking Faces: From Speech Perception to a Behavioral Principle*. MIT Press: Cambridge, MA, 1998.

Massaro, D.W. & M.M. Cohen. "Evaluation and integration of visual and auditory information in speech perception". *Journal of Experimental Psychology: Human Perception and Performance, 9*, 753-771, 1983.

Massaro, D.W. & M.M. Cohen. "Perception of synthesized audible and visible speech". *Psychological Science, 1*, 55-63, 1990.

Massaro, D.W. & M.M. Cohen. "Speech Perception in Perceivers with Hearing Loss: Synergy of Multiple Modalities". *Journal of Speech, Language, and Hearing Research,* 42: 21-41, 1999.

Massaro, D.W. & P.B. Egan. "Perceiving affect from the voice and the face". *Psychonomic Bulletin and Review, 3*, 215-221, 1996.

Massaro, D.W. & D. Friedman. "Models of integration given multiple sources of information", *Psychological Review, 97(2)*, 225-252, 1990.

Massaro, D.W. & D.G. Stork. "Speech recognition and sensory integration". *American Scientist, 86*, 236-244, 1998.

Massaro, D.W., M.M. Cohen & P.M.T. Smeele. "Cross-linguistic Comparisons in the Integration of Visual and Auditory Speech," *Memory and Cognition, 23,(1)* 113-131, 1995.

Massaro, D.W., M.M. Cohen & L.A. Thompson. "Visible language in speech perception: Lipreading and reading," *Visible Language, 22*, 9-31, 1988.

Massaro, D.W., M.M. Cohen, C.S. Campbell & T. Rodriguez. "Bayes factor of model selection validates FLMP". *Psychonomic Bulletin & Review*, 8, 1-17, 2001.

Massaro, D.W., M. Tsuzaki, M.M. Cohen, A. Gesi & R. Heredia. "Bimodal Speech Perception: An Examination across Languages", *Journal of Phonetics, 21*, 445-478, 1993.

Mattingly. I.G. & M. Studdert-Kennedy, (Eds). *Modularity and the motor theory of speech perception*. Hillsdale, NJ: Lawrence Erlbaum, 1991.

McGurk, H. & J. MacDonald. "Hearing lips and seeing voices". *Nature, 264*, 746-748, 1976.

Munhall, K.G. & Y. Tohkura. "Audiovisual gating and the time course of speech perception". *Journal of the Acoustical Society of America, 104*, 530-539, 1998.

Myung, I.J. & M.A. Pitt. "Applying Occam's razor in modeling cognition: A Bayesian approach". *Psychonomic Bulletin & Review, 4*, 79-95, 1997.

Oerlemans, M. & P. Blamey. "Touch and auditory-visual speech perception". In: Campbell, R., B. Dodd, & D. Burnham (Eds), *Hearing by Eye II* (pp. 267-281). East Sussex, UK: Psychology Press Ltd, 1998.

Palmer, S.E. *Vision Science: Protons to Phenomenology*. Cambridge, MA: MIT Press, 1999.

Pitt, M.A. & J. M. McQueen. "Is Compensation for Coarticulation Mediated by the Lexicon?" *Journal of Memory and Language, 39*, 347-370, 1998.

Rosenblum, L.D. & H.M. Saldana. "An audio-visual test of kinematic primitives for visual speech perception". *Journal of Experimental Psychology: Human Perception and Performance,* 22, 318-331, 1996.

Rosenblum, L.D. & H.M. Saldana. "Time-varying information for visual speech perception". In: Campbell, R., B. Dodd, & D. Burnham (Eds), *Hearing by Eye II* (pp.61-81). East Sussex, UK: Psychology Press Ltd, 1998.

Schindler, R.A. & M.M. Merzenich. *Cochlear Implants*. New York: Raven, 1985.

Schwartz, J., J. Robert-Ribes, & P. Escudier."Ten years after Summerfield: A taxonomy of models for audio-visual fusion in speech perception". In: Campbell, R., B. Dodd & D. Burnham (Eds), *Hearing by Eye II* (pp. 85-108). East Sussex, UK: Psychology Press Ltd, 1998.

Sekiyama, K. "Face or voice? Determinant of compellingness to the McGurk effect". *Proceedings of AVSP'98*. Terrigal – Sydney, Australia, 1998.

Tyler, R.S., J.M. Opie, H. Fryauf-Bertschy & B.J. Gantz. "Future directions for cochlear implants". *Journal of Speech-Language Pathology and Audiology, 16*, 151-164, 1992.

Warren, R.M. "Perceptual restoration of missing speech sounds". *Science, 167*, 392-393, 1970.

16. AFFILIATION

Dr. Dominic W. Massaro
Department of Psychology
University of California
Santa Cruz, CA 95064 USA
URL: http://mambo.ucsc.edu/psl/dwm/

17. ACKNOWLEDGEMENT

The research and writing of the paper were supported by grants from National Science Foundation (Grant No. CDA-9726363, Grant No. BCS-9905176, Grant No. IIS-0086107), Public Health Service (Grant No. PHS R01 DC00236), Intel Corporation, the University of California Digital Media Program, and the University of California, Santa Cruz.

A. D. N. EDWARDS

MULTIMODAL INTERACTION AND PEOPLE WITH DISABILITIES

1. MODALITIES, COMPUTERS AND PEOPLE WITH DISABILITIES

What is the connection between computers, multiple modalities and people with disabilities? A traditionally scientific chapter might start out with definitions of these terms. However, there is problem in doing that in this case, which is that only one of them – 'computer' – is at all easy to define.

Daily activities of people take the form of interactions between them and their environment (where 'environment' is meant in a broad sense, encompassing other people as well as the physical surroundings). These interactions occur through an 'interface' which uses the physical, cognitive and sensory functions of the person. If any of those functions is impaired to the extent that the person finds forms of interaction difficult or impossible, then that person is said to be disabled (UN, 1981). The degree to which that disability handicaps the person depends on the extent to which the impaired function can be supported or substituted.

Before the discussion becomes too abstract, let us consider some examples. A person with a hearing impairment has difficulty interacting with the auditory component of the environment. A hearing aid may help them to continue to operate in an auditory mode, which amounts to supporting the impaired channel. Yet, if the impairment is so severe that the hearing aid cannot assist, then they may still take part in conversation by substituting non-auditory channels. That is to say that the visual channel can be used to pick up the visible cues of speech (lip and tongue movements, facial expressions *etc.*) and hence substitute for the auditory channel.

This example is a good one because it illustrates how compensation may take place at a human level (the physical mechanisms of speech production) or by the application of technology. Often people with disabilities can be accommodated within human interactions because of the richness of the interaction, but where they cannot, technology has an increasing role to play – and that improvement is largely due to the broadening of the technology to exploit more modes of interaction.

The excitement about the technology in this area is that it is opening new opportunities. The technology can make some tasks easier for people with disabilities and in many cases can make things possible that were previously impossible. This chapter will describe a number of examples of this. It is the extension of the technology into new modalities of interaction that is making new possibilities viable.

Speech-reading (the current, more accurate term for 'lip-reading') is an example of a mapping from one communication channel to another, in this case from the auditory to the visual. The availability of multimodal technology facilitates this kind

B. Granström et al. (eds.), Multimodality in Language and Speech Systems, 73–92.

of mapping with the aid of technology. That is to say that speech-reading makes use of inherent redundancy in speech communication, but where such redundancy is not present, such mappings can be created technologically. This is why multimodal technology is such an important development for people with disabilities.

2. COMMUNICATION AND THE SENSES

Communication takes place via the five senses:

vision,
touch,
smell,
taste,
hearing.

Each of the sensory channels has particular characteristics, strengths and weaknesses. In many ways, vision is primary. A large proportion of the brain is devoted to visual processing and studies have shown (Mayes, 1992) that when conflicting signals are presented on the visual channel and another one, it is the visual one that will tend to be believed. Vision is very powerful, so that large amounts of information can be presented visually at any time and the real power of vision comes from the fact that it is possible to focus attention very precisely. There may be many objects and events in any visual scene and the viewer may have the impression of taking in all that information at once. In truth, the field of attention is very narrow – but that attention can be switched very quickly. Thus an event in the periphery of vision will attract attention and the eyes will be shifted to focus on it. In this section, the primacy of vision in current human-computer interfaces is discussed as well as the possibility of shifting some communication to the other senses, where appropriate.

Touch is an interesting case. In some ways it is an under-regarded form of communication. The only formal tactile languages are those used by blind people. Braille is the best-known one, but there is also the less-known Moon language[1]. While sighted people may think that their use of tactile communication is negligible touch-typing has become a major component of communication in this computer-oriented age. The majority of computer users are untrained and cannot truly touch type, but nevertheless they do rely on tactile feedback as part of their typing activity. There are also many other situations in which people rely largely on tactile feedback in interacting with switches, buttons and such-like (e.g. secondary controls in a car, such as heating and radio switches, which are activated without diverting visual attention from the road, the primary task).

The tactile senses are generally not only associated with pressure and feedback from physical contact, but also with sensations of temperature. This has been

---

[1] *Moon is used almost exclusively in the UK. It is based on tactile shapes that are more akin to printed letters and therefore more easily learned by people who have lost their sight later in life, after having had experience of visual reading. (RNIB (1996). This is Moon, RNIB. http://www.rnib.org.uk/braille/moonc.htm.).*

suggested as the basis of a possible form of communication (e.g. Challis, Hankinson *et al.*, 1998) but (at least for the present) this is not practical, not the least because of health and safety considerations.

We usually associate touch with the cutaneous feedback from the skin (mainly of the fingers) in contact with objects. There is, however, another, related form of bodily feedback, usually referred to as kinaesthetic. That is the information that we have about our limbs and other body parts in terms of the awareness of our muscles. The combination of tactile and kinaesthetic can be referred to as *haptic* (Oakley, McGee *et al.*, 2000).

Smell is another important form of communication. The exact level to which people use it is disputable. There is clear evidence of its having a large influence on interactions between animals, but many people would prefer to suggest that human behaviour has risen above the influence of pheromones. Yet, even if smell does not play a part in inter-personal communication, it can carry some very important messages. People are generally very sensitive to smells as warnings: the presence of a fire, that food has gone off and such-like. Also, it is suggested that ambient smells can have an important effect on people's moods, which could turn out to be influential in commercial situations, such as consumer e-commerce web sites. Hitherto technology has not existed to control olfactory messages; it has not been possible to generate smells of particular types – though work is proceeding in that area (Youngblut, Johnson *et al.*, 1996).

Taste is very closely related to smell. In fact, taste is quite a crude sense and most of the sensations that we attribute to taste are in fact the results of the smells accompanying them. We can only distinguish four primary tastes: sweet, bitter, salty and sour and the richer sensations that we derive when we drink a glass of wine, for instance, are in fact generated by the aroma of the wine in the glass just below our nose. Like smell, it is not really possible to generate tastes on demand and there is the further complication that to be tasted a substance must come in contact with the inside of the mouth – which raises a wide range of health and safety considerations!

Technological constraints imply that in technology-mediated communication it is practical to use the senses of vision, hearing and touch. Physical, sensory and cognitive impairments may mean that one or more of these senses is unavailable or inefficient. It is the role of technology to supplement or replace the lacking function. Taking one form of information to make it accessible via a different channel implies a *mapping*. That is a main theme of this chapter – the technological facility to map information between different modalities in order to accommodate the needs of users with disabilities.

It may be said that the designers of modern computer interfaces exploit the power of vision, in making maximum use of visual displays. Another view would be that such designers are lazy; if more information is required, they will slap another 'widget' onto the display, so leaving it to the user to cope with this extra complexity. With more thought, there might be better ways of presenting the new information, ways that will not increase the visual complexity and the user's task load. It is to be expected (and hoped) that in the future designers will be aware of the possibility of using different channels when appropriate. For instance, Brewster (Brewster, 1994) demonstrated that by analysing human-computer interfaces, in terms of *events, status* and *modes* it was possible to identify information that was hidden from the user, which then could be presented in an auditory form.

While sight is generally assumed to work well in processing simultaneous sources of information, hearing is usually assumed to not be good at such parallel processing. This is not necessarily true, though. Massively parallel information can be presented in sounds - if they are designed in the right way. Once again it is more a question of attention switching. Buxton's example (Buxton, 1989) is of driving a car, when the driver might be engaged in a conversation, but at the same time may have the radio on, be monitoring auditory signals from the car (turn indicator clicking, note of the engine *etc.*) and be aware of external signals, such as an ambulance siren. In the event of a significant change to the auditory scene (such as a traffic report on the radio or the onset of a 'clunking' noise in the engine), the driver may have to withdraw from the conversation to switch attention to the alternative event. Another popular example is known as the 'cocktail party effect'. In a busy room with conversations all around, it is possible to have a dialogue with another person without interference. Yet the auditory system still monitors the ambient sounds, so that, for instance, if the person's name is spoken by someone elsewhere in the room, their attention will be drawn to that and away from their current conversation.

This processing of different sources of sound is known as auditory streaming (Bregman, 1990) and is mentioned again in Section 5. One difference from visual attention is that sound is not directional, so that it is not possible to focus exclusively on one sound to the same extent. Hearing has a degree of directional discrimination, but this is not very precise in humans, who have their ears on the side of their heads and which cannot be turned independent of the head (unlike some animals). Thus it is a natural reaction to turn the head in the direction of a sound in order to locate it or to listen to it.

Another important difference with hearing is that sound is inherently transient; it exists in time. It is not possible to review or re-examine a sound. The only mechanisms for doing this are dependent on memory. For instance, eye tracking experiments show that the process of reading visual text is not a simple left-to-right serial scan, but involves moving back and forth, revising and reinforcing words read. By contrast, spoken words are lost as soon as they are spoken. All that remains is an internal representation, the form of which depends on the amount of information presented and on time. (See Chapter 2 of Pitt, 1996, for more details).

Working visually one has a broad field of view, but the ability to focus on a narrow portion of that input. By contrast, tactile communication is inevitably narrowly focused. That is assuming that tactile communication takes place through the fingertips, which are the most convenient means. The 'field of view' of the fingertips is very narrow, and it is not possible to build up a larger picture by moving the fingers around, in the way that visual pictures are build by rapid movements of the eyes. Use of the tactile senses can be improved by training, so that people can learn to some extent to build more complete pictures by tactile exploration.

The tactile sense has very low resolution. The number of different surface textures that can be recognized by most individuals is small. The number can be increased by using different materials (e.g. rubber, leather, paper and aluminium) but it is usually not practical to produce tactile materials (effectively collages) using such materials.

Impairments which affect one sense or communication channel can be alleviated by substitution of a different channel. That implies mapping information from one form to another. The above discussion has illustrated that the channels have different inherent characteristics. Hence such mappings are not always straightforward. Before we go on to examine such mappings, though, it is necessary to clarify a further point, which is that channels are not simple, uni-dimensional entities.

3. MODALITIES

It is important to realize that although here we are considering just three channels of communication, corresponding to the available senses, there are many more *modalities* – of communication[2].

As an example, there are a variety of forms of visual communication, and printed forms may themselves be subdivided into textual and pictorial. Mappings need not be only between channels, but may also be from one modality to another. For instance, textual, written communication is not usable by someone who is illiterate, but pictures may be. (See the examples of picture-based communication in Section 4.3).

Even within one modality, important variations exist. For instance, there is more than one style of writing; the full, emotional message of a poem is different from the dry, factual information within a technical manual.

In principle, the same information can be communicated in different modalities. In practice such mappings are not pure. That is to say that in translation to another modality, the meaning is usually altered, albeit subtly. For instance, speech and writing is based on words, but speech includes elements of intonation and prosody that are mostly lost in text. Thus, the simple utterance

(1) It's raining.

might be a statement, but if spoken with a certain intonation (a rising pitch, in British English), could be transformed into a question. A writer using those words as a question would signal the intention with a question mark, but there are other, more subtle variations that cannot be captured grammatically. For instance (an example borrowed from Stevens, 1996),

(2) Robert does research on drugs.

---

[2] *'Multimodality' is the topic of this book – and yet few authors agree exactly as to the meaning of the word modality. (See, for instance, the discussion in Blattner, M. and Dannenberg, R. B. (1992). Introduction: The trend toward multimedia interfaces. (in) Multimedia Interface Design. M Blattner and R. B. Dannenberg (Eds.), New York, ACM Press, Addison-Wesley: pp. xvii-xxv.). In despair of finding consensus on a definition, this chapter does not attempt any new definitions, but attempts to at least be self-consistent.*

might be read as meaning that Robert is a pharmaceutical investigator – or that he performs research work while under the influence of narcotics. Mapping between modalities is, therefore, not always as simple as one might hope. The text of Sentence (2) could easily be passed to a speech synthesizer, thereby mapping from text to speech, but the synthesizer would have to impose one or other of the interpretations – possibly the wrong one.

Table 1 is based on one produced by Jens Allwood showing how the different components of human dialogue are perceived by the 'listener'. Such a table might be used as a basis for remapping communication to accommodate a disabled communicator. For instance, for a deaf person, the *hearing* column is unavailable and so one of the other senses might be used instead. Vocabulary might be shifted into the Vision column, by use of text instead of speech.

Table 1. Modalities in human dialogue (after Allwood). Each row is an expressive modality while the columns indicate how the modality is perceived by the 'listener'.

		Hearing	*Vision*	*Touch*	*Smell*	*Taste*
Speech	Prosody/ phonology	3	3			
	Vocabulary	3				
	Grammar	3				
Gestures	Head movements		3			
	Facial gestures		3			
	Manual gestures		3	3		
	Body movement	3	3	3		
	Posture		3			
	Touch			3		
	Smell				3	
	Taste					3
	Writing		3			

It would be constructive if such a table could be extended to all forms of communication and used as the basis of (semi-) automatic remediation of communications impairments. However, this is not practical – at least not at the current state of knowledge. For a start we would need a fine-grained taxonomy of communication channels and acts. Whereas Table 1 demonstrates that psycholinguists have a good grasp of the nature of human dialogue, there are many other forms of communication that are less well understood. An extreme example would be a painting. Who can say exactly *what* the 'message' of (say) the Mona Lisa is, never mind how that message is communicated?

Table 1 also illustrates another point that should be borne in mind. It represents interpersonal dialogue and the very size of the table demonstrates the richness of that communication. However, most communication aid devices do not span the whole table. In other words, they tend to concentrate solely on the Speech rows of the table. The assumption is that it is sufficient to generate the vocabulary and grammar of dialogue, but that is to exclude all the other facets. Communication aids are discussed further in Section 4.3

4. EXAMPLES OF NOVEL MAPPINGS

In this section we will illustrate the potential to use technology to map between modalities in specific applications that are aimed at alleviating the affects of disabilities. These examples are of existing systems, some (commercially) available and some that are rather more experimental. To an extent they represent the state of the art, but only hint at the applications that may eventually be possible with further development of knowledge and the technology.

4.1 Screen Readers

The screen reader is probably the clearest example of cross-modality mapping technology. Blind people do not have access to visual information and computers are very visual in operation. A screen reader is software which captures the information that is on a computer screen such that it can be translated into a non-visual form. That form may be auditory (mainly synthetic speech, but also non-speech sounds) or tactile – braille.

One of the important features of the screen reader is that it works with standard applications. That is to say that the blind person can use the same word processor or database package as sighted colleagues do because of the addition of the adaptation on top of the application software.

A screen reader cannot simply 'dump' everything on the screen into speech – because of the differences between vision and hearing listed earlier. That is to say that a computer screen generally contains a large amount of information and the user focuses on the area of interest at any time, but if all the information were presented in speech it would be an unfocused babble to the user. The screen reader therefore must provide some form of control. That is to say that the software must provide the control that the visual system does for a sighted user. It must present only the amount of information that the user can cope with at a time – but also make it possible to access other information. For instance, on a visual screen a warning may be flashed in the periphery to attract the user's attention and that same warning will be equally valuable to the blind user.

The miss-match between the characteristics and capacity of vision and hearing lead to difficulties for the screen reader designer. The simplest approach would be to filter out information. However, it is not for the designer to decide what the blind user shall have access to. If information has been provided in the visual interface, it is there for a reason, and the blind user must be allowed access to it. The designer does have to decide, though, which is the most salient information that must be presented immediately to the user, and what is less important information that will not be presented in parallel, masking the primary message. This secondary information may be made more difficult to access (i.e. in response to an explicit request from the user), but still present. The problem with this approach is that it complicates the interface between the user and the screen reader. This compounds the usability problem, because the screen reader in itself represents an addition to the complexity of the application being used. There is a difficult trade-off then between giving users as much power as possible but to do so in a way that they can utilize it, avoiding making it too complex to use. Human-computer interaction principles of simplicity, consistency and so on become even more critical. (Such principles are exemplified by guidelines such as Smith & Mosier, 1984, though they tend to be visually oriented).

In the early days of interactive computers, once control mechanisms had been devised, the problem of the screen reader designer was relatively simple – because the main mode of display was textual. In other words, the mapping from text to speech was easily achieved, through speech synthesis technology (Edwards, 1991). However, the advent of the graphical user interface (GUI) raised rather more complex problems. Now visual objects and properties other than text were significant and it was rather more difficult to convert these into non-visual forms.

The history of approaches to this problem is presented in Edwards (1995) and eventually resulted in commercially available screen readers for common GUIs. Some of the design questions and trade-offs are documented in Mynatt & Weber (1994).

4.2 Non-Visual Diagrams

While GUIs contain graphical elements, a particular problem for blind people is access to graphics in general, to diagrams and pictures. A number of approaches have been investigated experimentally, but none has yet been adopted as practical.

One potential mapping is into a tactile form. As discussed below, this can be achieved quite readily for static pictures; it is not practical for dynamic, changing graphics. The use of (computer) screens to display graphics is increasing and so it would be attractive to have some tactile counterpart: a tactile screen that can quickly switch between different pictures, or even display animations. Many developers have thought of this idea – but none has been able to find a practical solution.

Most of the devices are based on moving pins. A pin-raised proud of the others can be felt. A row of such pins becomes a tactile line and so on. The practical problems are caused by the fact that the device is electro-mechanical and hence prone to physical and mechanical problems. To achieve good resolution, the pins must be small and hence must be manufactured to high precision, making manufacture very expensive. Reliability is a problem. A mechanical fault may cause a pin to stick that may render the whole device practically useless. Refresh rates will be slow too because of mechanical delays.

A commercially available pin-matrix device is available from Metec, in Germany, and has the following features. It consists of 7200 pins (120 x 60) in a 37.2 x 18.6 cm rectangle. That gives a resolution of 3.2 pins per cm or 8 'dots' per inch. The device can be used to display braille, giving 15 lines of 6-dot braille cells. Pins are individually addressable, which means that software can be optimized to reduce the area refreshed. This is important because it takes of the order of 21.6 seconds to refresh the whole display. The price of one of these devices is of the order of EUR 60,000.

Static tactile representations are rather more easy to produce, and modern computing and printing technology has a role to play. The simplest way to produce tactile graphics is using what is known as *swell paper*. This is paper with a plastic coating. The plastic is heat sensitive such that it expands on heating. So, diagrams can be produced by photocopying a black-and-white picture onto the paper and then putting the paper through a heating machine. The black areas absorb more heat and swell to make a raised area.

The development of this technology has made it almost as easy (and cheap) to produce tactile diagrams as visual ones and some guidance exists as to how best to design such diagrams (e.g. Hinton, 1996). However, it is apparent that many designers have not appreciated the fundamental differences between sight and touch and the implications for good design of tactile diagrams. Often tactile diagrams are simply visual diagrams printed on swell paper. They may appear comprehensible and attractive to visual inspection, but may be of little use to the blind person. Challis (2000) and Challis & Edwards (2000) is developing guidelines for tactile diagram design that will avoid these errors.

Examples of the kind of error that can be made in this kind of visual to tactile mapping are the use of blank space. A plain white region of a visual map can meaningfully depict an empty area. However, if that same map is printed on swell paper and the reader places a finger on that blank area then no information is conveyed. There is no indication of what that area depicts or of what direction to move the finger in order to find useful information. That kind of feedback could have been provided if there had been some form of texture in that area.

Challis's work seems to suggest that good, effective tactile diagrams may be difficult to interpret visually. For instance, he has been developing a tactile representation of music for blind musicians, and Figure 1 shows his representation of the score shown in conventional notation in Figure 2.

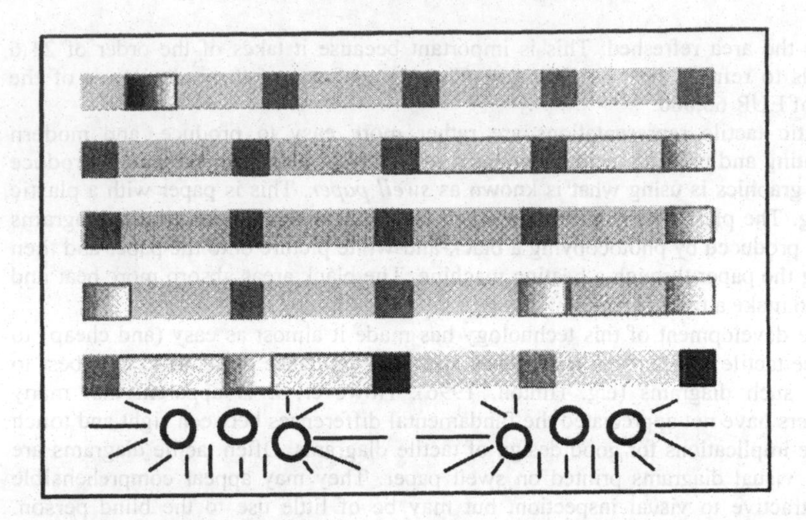

Figure 1. A Weasel overlay, representing the piece of music shown in Figure 2.

Figure 2. Common music notation representing the same piece of music as in Figure 1.

Sounds have also been used as a means of representing non-visual diagrams. Bennett & Edwards (1998) and Bennett (1999) experimented with the representation of simple 'box and line' diagrams (Figure 3) using speech and the non-speech sounds known as earcons (Blattner *et al.*, 1989; Brewster *et al.*, 1995). Rigas (1996) and Rigas & Alty (1997) developed Audiograph, which represented the shape and position of elements of diagrams using musical encodings. The confusingly named Audiograf (Kennel, 1996) is a prototype reader for diagrams that have hierarchies and connections. It is based on a model of audio-tactile exploration that assumes that there is a cycle of user interaction with the interface, through which the user enhances knowledge with regards to areas of uncertainty or interest by interrogating the system. In this way, information is incorporated into the user's knowledge base.

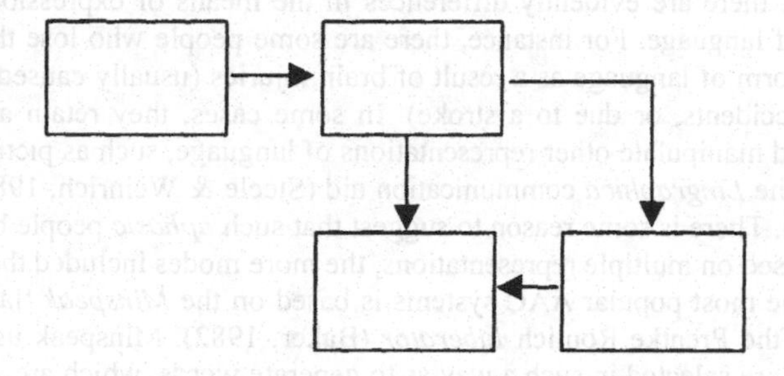

Figure 3. A simple box-and-line diagram that might be converted into an auditory form following the approach of Bennett.

Examination of pictures is one problem, but beyond that it would be most valuable if blind people could create their own graphics. Kurze (1996) has tackled this through a multi-modal approach, using both auditory and tactile feedback. Pictures are created using a heated stylus on swell paper, so that a tactile picture is created but at the same time audio feedback is generated.

A different approach to cross-modality mapping has been taken by Blenkhorn and colleagues (1994) to make software engineering graphical notations accessible to blind programmers. Dataflow diagrams are a graphical representation of software. There is an alternative, text-based representation in which components and the links between them are transformed into a matrix, known as an N-squared chart. This representation contains the same information, but it cannot be said to be perceptually equivalent. That is to say that there is greater cognitive effort on the user to extract the same information from the N-squared diagram. Generally sighted programmers would prefer the graphical representation over the N-squared diagram. However, the latter is easier to transform into a usable non-visual form, based on text to speech translation that is the basis of the *Kevin* tool (Blenkhorn & Evans, 1994).

4.3 Communication Aids

The use and generation of language is an important human characteristic (even if experiments with chimpanzees, dolphins and other animals may suggest it is not uniquely human). According to models such as that of Patterson & Shewell (1987) the mechanisms of production of written and spoken language are intimately related. For instance, they suggest that a writer 'hears' an utterance internally before writing it on paper. Therefore, it is a moot point as to whether communication via an external device (usually referred to as *alternative and augmentative communication,* or AAC) involves any mapping between modalities.

However, there are evidently differences in the means of expression and comprehension of language. For instance, there are some people who lose the ability to use written form of language as a result of brain injuries (usually caused by trauma, as in road accidents, or due to a stroke). In some cases, they retain an ability to recognize and manipulate other representations of language, such as pictures. This is the basis of the *Lingraphica* communication aid (Steele & Weinrich, 1986; Sacks & Steele, 1993). There is some reason to suggest that such *aphasic* people benefit from languages based on multiple representations, the more modes included the better.

One of the most popular AAC systems is based on the *Minspeak* 'language'[3] as available on the Prentke Romich *Liberator* (Baker, 1982). Minspeak uses pictorial symbols that are selected in such a way as to generate words, which are spoken with a synthetic voice – and can be displayed visually as text. Minspeak can operate at different levels. At its simplest, each picture represents a single word or phrase and utterances are created by stringing together a set of these components. There are 128 pictures to choose from. Such a limited range of utterances might be sufficient for a new user of the system who has a limited vocabulary, but for every-day speech a much wider vocabulary is required. This can be achieved by allowing multiple selections of pictures. That is to say, if two selections are required to select a word, then the vocabulary rises to $128^2 = 16,384$ words. In practice, the device may be programmed such that commonly used words and phrases are available through a single selection, while multiple selections are required for less frequently used ones.

In order to use Minspeak, the user must, of course, learn the mapping from sequences of pictures to language utterances. The Liberator can be programmed entirely by the user (or more likely their teacher or speech therapist). This means that it can be matched to the abilities of the individual. For some users it may be appropriate to use pictures that are accurate representations of real-world objects to generate the name of that object. For instance, a photograph of the user's mother could be used to generate the word 'mother'. This is the simplest form of mapping. However, to extend the range of utterances available, the pictures become more abstract and more generalized. The way that this is achieved in Minspeak is through the concept of 'semantic compaction'. That is to say the attachment of as many associations as possible to each picture. As is illustrated in the example below, this

[3] *Some people, e.g. Strong, G. W. (1995) argue that Minspeak is not truly a language in that it does not have a defined syntax and semantics.*

often relies on the use of puns, which evidently assist in memorization. Selection is also structured, so that choosing one symbol will set a theme, while the second selection will then specify the particular object or concept within that theme.

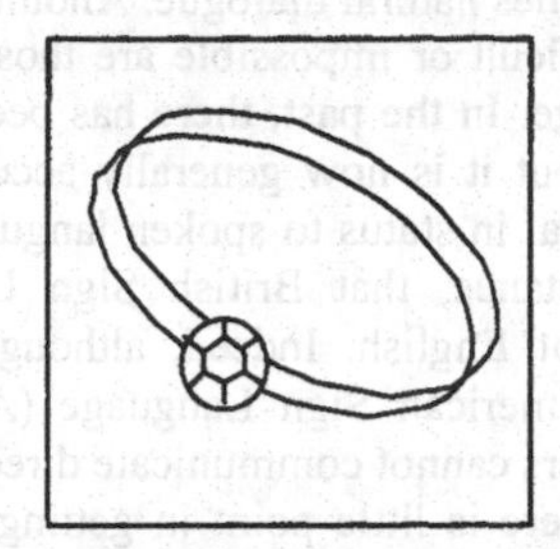

Figure 4. Icon-like pictures that might be used in Minspeak. The watch might be associated with a number of time-related concepts, while the ring might be associated with circular concepts (such as the number 0) as well as items which are expensive – associated with diamond rings.

For instance Figure 4 shows two pictures that might appear on a Liberator. The watch might be used for concepts associated with time. Selecting it twice could represent 'today', which is easier to remember if one thinks of 'two-day'. The ring can be associated with circular objects (including the number 0). The word 'tomorrow' could be watch + ring. That could be explained – and remembered – as watch setting the time theme and then the ring is reminiscent of the three letter Os in the word tOmOrrOw. A more detailed example of generating an utterance using Minspeak is available in Chapter 4 of Edwards (1991). Skilled Minspeak users can generate speech as quite fast rates and it would be interesting to investigate the way they map internally from symbols to speech and to see whether the process maps into the Patterson & Shewell (*op. cit.*) model of language production.

Since Minspeak relies on selection of symbols, a variety of different techniques can be used to make those selections. The obvious one is to press keys. Other users may find it easier to select by pointing at the pictures using an infra-red source which can be attached to whichever part of the body they can best control (e.g. the head). Or, finally, selection may be made using a scanning system, so that a picture is selected by pressing a switch when it is highlighted.

One limitation of current communication aids becomes apparent when one considers them within a multimodal context. That is that designs concentrate solely on the verbal content of inter-personal communication. Linguists such as Allwood recognize that there is more to it than that (as exemplified in Table 1), but AAC designers tend to think of the task as being a simple translation from one modality of communication into speech. That is to exclude the other important modalities (bodily gestures, facial gestures, non-verbal utterances *etc.*) as if they are mere decorations to accompany the speech, when in fact they carry a lot of information. Researchers at Dundee University (Newell, Arnott *et al.*, 1995) are an exception in that they have looked at means of including non-verbal utterances in augmented communications.

The commonest mode of output of AAC devices is synthetic speech, which most closely matches natural dialogue. Another group of people who find spoken communication difficult or impossible are those who are deaf. Many such people rely on sign language. In the past, there has been some controversy over the status of sign languages, but it is now generally accepted that they are true languages. Though they are equal in status to spoken languages, they do not resemble them. That is to say, for instance, that British Sign Language (BSL) is not a word-to-gesture translation of English. Indeed, although Britons and Americans speak the same language, American Sign Language (ASL) is entirely different from BSL; deaf British signers cannot communicate directly with Americans.

Again there is little point in getting into a debate as to the modes of communication involved, but it is apparent that technology may have a role in bridging the communication gaps that exist between signers and others (i.e. the vast majority of the hearing population). Research is under way in a number of institutions into the possibility of automatic sign language to speech translation. However, Edwards (1998) suggests that there is a long way to go before this will be achieved. Translation in the other direction, from text (or speech, via voice recognition) into sign (rendered by an animated, cartoon-like avatar) should also become possible.

Multimodal technology is already playing a role, though, in other ways. Hitherto a problem with sign language is that it has not been easy to capture it in disembodied formats. That is to say that whereas (most of) speech can be translated into written text, there has been no equivalent for sign. Paper sign language dictionaries have relied on static photographs, but, because signs are often dynamic, these have had to be augmented by arrows and similar notations. Now, with the ready availability of video, properly expressive dictionaries are possible and these are easily made available over the web. (See, Lapiak, for instance).

Speech recognition technology has greatly improved in recent years. Now recognition of connected speech is available at a low price. The recognition levels are often still frustratingly low for many users, but they are good enough as an alternative to the keyboard for many people who cannot use the keyboard for one reason or another. This would include people with physical impairments, which preclude them from using a keyboard. Ironically a growing cause of such impairments is the increasing use of information technology, leading to repetitive strain injuries (RSI[4]) caused by typing.

An intriguing possibility for the exploitation of translation between modalities, is the combination of speech-to-text and text-to-speech which might assist communication for people with speech impairments. Some (speaker-dependent) speech recognition techniques rely more on consistency of the speaker than conformity to recognized language. Therefore, such a system can be trained that a particular sound has a specified meaning, regardless of whether that sound resembles the pronunciation of the word – as long as that sound is consistently reproduced by the speaker.

---

4 *At present a variety of different terms are being used to describe this phenomenon. Another term is 'work-related upper limb disorders' or WRULD.*

Thus, a person who speaks in an idiosyncratic manner that (untrained) listeners find hard to understand, might speak to a machine that would translate into synthetic speech which is easy to comprehend. Experiments have been carried out which suggest that this approach would be viable (Edwards & Blore, 1995; Schulz & Wilhelm, 1992), though it has yet to be implemented in practice.

Speech technology may have a particular role for the large group of people who have problems with written language, due to dyslexia. For those who have difficulty writing, speech input may be a useful alternative. This is being used increasingly, though Elkind & Shrager (1995) demonstrated that this may not be as efficient an alternative as is assumed. The problem is that speech recognizers are less than 100% accurate. In case of doubt, the system will present the user with a menu of possible interpretations (near homophones) of the word spoken. Yet, what task is a dyslexic person likely to find more difficult than recognizing the correct spelling of a word?

Speech output has a role, though. Many dyslexics find that they can spot errors when their text is read out to them (even in a synthetic voice) that they would not spot if they read the text themselves.

4.4 Further Examples

There is quite a large range of assistive devices available for people with disabilities, which rely on multiple modalities. This list will only increase in size as the technology develops. Some more of those currently available are described in this section.

4.4.1. Optacon

Braille is one example of translation of (visual) text into a tactile form. Modern technology for optical character recognition, translation and printing (embossing) can facilitate the mapping, but not in real time. The Optacon is a device, which achieves a visual-to-tactile translation automatically as the user reads. A miniature television camera is scanned across a printed page and the shape of the letter under the camera is reproduced on a pad of vibrating pins. The user holds a finger on that pad and can feel the shape of the letter. In this way, a (trained) blind user can read printed text without assistance.

Although it has been a successful device and very popular with its users, the Optacon is no longer manufactured.

4.4.2. Speech viewers

Deaf people who try to learn to communicate orally have difficulty in speaking because they cannot hear their own voices well. Their training can be supported by the use of device, which will present them with a visual representation of their speech sounds. Typically the speech therapist or teacher will speak into the device, producing a sample visual representation and then the deaf person will attempt to produce a similar pattern. Alternatively this approach may take the form of a game. The deaf person (usually a child) will achieve a good score by producing vocal sounds of the appropriate kind. IBM's SpeechViewer is an example of such a device.

4.4.3. Text-to-sign

People who are both deaf and blind effectively only have their tactile senses for receiving communication. They may communicate with a signer by feeling the person's hands. For people who do not know sign language, experimental technology – known as *Ralph* (Jaffe, 1994) – has been developed whereby a robot hand will perform the signs of finger spelling. Thus, the non-signer can type input that is displayed on the robot hand which is felt by the deaf blind person.

4.4.4. Sound graphs

Graphs are a very powerful visual representation of quantitative information. Many of the same properties can be captured in a representation in sounds. Usually the x-axis is represented by time and the height of a curve on the y-axis by the pitch of a note. One of the first publications of this idea is Mansur *et al.*, (1985), one implementation is described in Edwards & Stevens (1993) and a version that is currently available commercially is the Audio Graphing Calculator (AGC), available from *ViewPlus* (http://www.ViewPlusSoft.com). An evaluation of the usefulness of sound graphs can be found in Bonebright *et al.* (2001).

5. A METHODOLOGY FOR MULTIMODAL DESIGN

The above discussion should have established the proposition that multimodal technology has a valuable role to play in the remediation of some of the effects of disabilities. A remaining question, though, is how best to make use of this opportunity. That is to say that there is an existing fount of knowledge as to how to design conventional, limited-modality interfaces and systems, but there is less known about how to design a good multimodal system. This is the question that has been addressed by Mitsopoulos (2000) in his development of a methodology for multimodal design.

It is beyond the scope of this chapter to outline the methodology in full, but it will be outlined here. Full details can be found instead in Mitsopoulos (2000), while different aspects are described and examples worked through in Mitsopoulos & Edwards (1998); Mitsopoulos & Edwards (1999a); Mitsopoulos & Edwards (1999b). The motivation behind the methodology is to support the designer of a multimodal interface. It does not present simple guidelines but steps that the designer can go through. The methodology has been applied to the transformation of visual computer interfaces into auditory counterparts, but should be equally applicable to other modalities and to designing interfaces from scratch as well as adaptation.

The methodology works at three levels:

- conceptual
- structural
- implementation

At the conceptual level, the designer must consider the *tasks* to be undertaken, in terms of the information required by the user. That information can be classified according to the dimensions suggested by Zhang (1996): *nominal, ordinal* and *ratio*.

Having thus ascertained the information required to support the tasks, the structure that will support the tasks can be identified. The structural components of an auditory interface are *streams* as described in Bregman (1990). These are the auditory counterparts of the *objects* that make up any visual scene.

Now, having designed the structure, it is necessary to choose components that will implement it. In the case of an auditory interface, that means choosing sounds that will be perceived as belonging to the streams identified. Cognitive models, such as Interacting Cognitive Subsystems (ICS, Barnard & May, 1994) can give some guidance at this level – though the auditory and tactile contribution to cognition is less well developed in this model as yet.

6. CONCLUSIONS

The world is a sensually rich environment. That humans have thrived within it is to a significant extent due to the ability of the senses to capture information and of cognition to process that information. The richness of the environment enhances communication as it allows for overlap and redundancy.

The senses are not omnipotent, though; we cannot hear the whole range of auditory frequencies nor see the whole electromagnetic spectrum, and we do not necessarily understand all that we can perceive. Technology has enabled us to extend our inherent capabilities, though. For instance, we use the communication properties of radio waves by transforming them to and from signals we can perceive.

Some people's window onto the world is further restricted as a result of physical, sensory or cognitive impairments – to the extent that they are said to be disabled, but once again technology has a role to play. It can again broaden that window so as to reduce the deleterious effect of the impairment. To do this most effectively, we need to imitate nature and to ensure that as many modes of communication are recruited as possible, in such a way that they reinforce and support each other. It is only recently, with development of digital information and communication technology (ICT), that this has become a realistic proposition. This introduces new and exciting possibilities. It is likely that technology will redefine what we mean by 'disability'. Most people who wear spectacles or eyeglasses would not class themselves as visually disabled, though without the glasses they would find common tasks difficult. So it may be that in the future people who rely on multimodal ICT will be able to operate just as efficiently as their unaided peers and so not be classed or categorized as any different.

7. REFERENCES

ASL Dictionary Online. http://www.bewellnet.com/dario/asl_dictionary_online_practical.htm.

Baker, B. Minspeak. *Byte* 7(9): pp. 186–202, 1982.

Barnard, P. & J. May. Interactions with Advanced Graphical Interfaces and the Deployment of Latent Human Knowledge. In: *Interactive Systems: Design, Specification and Verification.* F. Paterno (Ed.) Heidelberg, Springer-Verlag, 1994.

Bennett, D. *Presenting diagrams in sounds for blind people*, DPhil thesis, University of York, Department of Computer Science, 1999.

Bennett, D.J. & A.D.N Edwards. Exploration of non-seen diagrams. In: *Proceedings of ICAD '98 (International Conference on Auditory Display)*, S.A. Brewster & A.D.N. Edwards (Eds.), Glasgow: British Computer Society, 1998.

Blattner, M. & R.B. Dannenberg. Introduction: The trend toward multimedia interfaces. In: *Multimedia Interface Design*. M. Blattner & R.B. Dannenberg (Eds.), New York: ACM Press, Addison-Wesley: pp. xvii-xxv, 1992.

Blattner, M.M., D.A Sumikawa & R.M. Greenberg. Earcons and icons: Their structure and common design principles. *Human-computer Interaction* 4(1): pp. 11–44, 1989.

Blenkhorn, P. & D.G. Evans. A method to access computer aided software engineering (CASE) tools for blind software engineers. In: *Computers for Handicapped Persons: Proceedings of the 4th International Conference, ICCHP '94*, W.L. Zagler (Ed.), pp. 321-328, Springer-Verlag, 1994.

Bonebright, T.L., M.A. Nees, T.T. Connerley & R, M.C.G. Testing the effectiveness of sonified graphs for education: A programmatic research project. In: *ICAD 2001*, J. Hiipakka, N. Zacharov & T. Takala (Eds.), Espoo, Finland, Helsinki University of Technology, 2001.

Bregman, A.S. *Auditory Scene Analysis*. Cambridge, Massachusetts: MIT Press. 1990.

Brewster, S.A. *Providing a structured method for integrating non-speech audio into human-computer interfaces*, DPhil Thesis, University of York, Department of Computer Science, 1994.

Brewster, S.A., P.C. Wright & A.D.N. Edwards. Experimentally derived guidelines for the creation of earcons. In: *Adjunct Proceedings of HCI'95: People and Computers*, G. Allen, J. Wilkinson & P. Wright (Eds.) pp. 155–159, Huddersfield, British Computer Society. 1995.

Buxton, W. Introduction to this special issue on nonspeech audio. *Human-Computer Interaction* 4(1): pp. 1-10, 1989.

Challis, B.P. *Design principles for tactile communication within the human-computer interface*, DPhil thesis, University of York, Department of Computer Science, 2000.

Challis, B.P. & A.D.N. Edwards. Design principles for tactile interaction. In: *First International Workshop on Haptical Human-Computer Interaction*, S. Brewster (Ed.) pp. 98-101, Glasgow, British Computer Society. 2000.

Challis, B., J. Hankinson, T. Evreinova & G. Evreinov. Alternative textured display. In: *Computers and Assistive Technology, ICCHP '98: Proceedings of the XV IFIP World Computer Congress*, A.D.N. Edwards, A. Arato & W.L. Zagler (Eds.) pp. 37–48, Vienna & Budapest, Austrian Computer Society. 1998.

DigiScents. *Digiscents: A revolution of the senses*. http://www.digiscents.com/. 2000

Edwards, A.D.N. *Speech Synthesis: Technology for Disabled People*. London: Paul Chapman. 1991.

Edwards, A.D.N. The rise of the graphical user interface. *Information Technology and Disabilities* 2(4), (http://www.isc.rit.edu/~easi/itd/itdv02n4/article3.html). 1995.

Edwards, A.D.N. Progress in sign language recognition. In: *Gesture and Sign Language in Human-Computer Interaction*. I. Wachsmuth & M. Frölich (Eds.), Berlin: Springer: pp. 13–21. 1998.

Edwards, A.D.N. & A. Blore. Speech input for persons with speech impairments. *Journal of Microcomputer Applications* 18: pp. 327–333, 1995.

Edwards, A.D.N. & R.D. Stevens. Mathematical representations: Graphs, curves and formulas. In: *Non-Visual Human-Computer Interactions: Prospects for the visually handicapped*. D. Burger & J.-C. Sperandio (Eds.), Paris, John Libbey Eurotext: pp. 181-194, 1993.

Elkind, J. & J. Shrager. Modeling and analysis of dyslexic writing using speech and other modalities. In: *Extra-ordinary Human-Computer Interaction: Interfaces for Users with Disabilities*. A.D.N. Edwards (Ed.) New York: Cambridge University Press: pp. 145–168, 1995.

Hinton, R. *Tactile Graphics in Education*. Edinburgh: Moray House Publications, 1996.

Jaffe, D.L. *Ralph: A fourth generation fingerspelling hand*. http://guide.stanford.edu/Publications/-dev2.html. 1994.

Kennel, A.R. Audiograf: A diagram reader for the blind. In: *Proceedings of Assets '96*, pp. 51–56, Vancouver, ACM, 1996.

Kurze, M. TDraw: A computer-based tactile drawing tool for blind people. In: *Proceedings of Assets '96*, pp. 131–138, Vancouver, ACM, 1996.

Lapiak, J.A. Handspeak: A sign language dictionary online, http://dww.deafworldweb.org/asl/

Mansur, D.L., M. Blattner & K. Joy. Sound-Graphs: A numerical data analysis method for the blind. *Journal of Medical Systems* 9: pp. 163-174, 1985.

Mayes, T. The 'M' word: Multimedia interfaces and their role in interactive learning systems. In: *Multimedia Interface Design in Education*. A. D. N. Edwards and S. Holland (Eds.), Berlin, Springer-Verlag. 76: pp. 1-22. 1992.

Mitsopoulos, E. *A principled approach to the design of auditory interaction on the non-visual user interface*, DPhil thesis, University of York, Department of Computer Science, 2000.

Mitsopoulos, E.N. & A.D.N. Edwards. A methodology for the specification of non-visual widgets. In: *Adjunct Conference Proceedings of HCI International '99*, H.-J. Bullinger and P.H. Vossen (Eds.) pp. 59-60, 1999a.

Mitsopoulos, E.N. & A.D.N. Edwards. A principled design methodology for auditory interaction. In: *Proceedings of Interact 99*, M. A. Sasse & C. Johnson (Eds.) pp. 263-271, Edinburgh, IOS Press. 1999b.

Mitsopoulos, E.N. & A.D.N. Edwards. A principled methodology for the specification and design of non-visual widgets. In: *Proceedings of ICAD '98 (International Conference on Auditory Display)*, S. A. Brewster & A.D.N. Edwards (Eds.), Glasgow, British Computer Society. 1998.

Mynatt, E.D. & Weber, G. Nonvisual presentation of graphical user interfaces: Contrasting two approaches. In: *Celebrating Interdependence: Proceedings of Chi '94*, C. Plaisant (Ed.) pp. 166-172, Boston, New York: ACM Press. 1994.

Newell, A.F., J. L. Arnott, *et al.* Intelligent systems for speech and language impaired people: A portfolio of research. In: *Extra-Ordinary Human-Computer Interaction: Interfaces for Users with Disabilities*. A.D.N. Edwards (Ed.) New York, Cambridge University Press: pp. 8–102. 1995.

Oakley, I., M.R. McGee, S.A. Brewster & P.D. Gray. Putting the feel in 'look and feel'. In: *The Future is Here: Proceedings of Chi 2000*, T. Turner, G. Szwillus, M. Czerwiniski & F. Paternò (Eds.) pp. 415-422, The Hague, NL, ACM Press Addison-Wesley, 2000.

Patterson, K. & C. Shewell. Speak and spell: Dissociations and word-class effects. In: *The Cognitive Neuropsychology of Language*. G.S. Max-Coltheart & R. Job (Eds.), London: Lawrence Erlbaum Associates: pp. 273–294, 1987.

Pitt, I.J. *The Principled Design of Speech-Based Interfaces*. DPhil Thesis, University of York, 1996.

Rigas, D. *Guidelines for auditory interface design: An empirical investigation*, unpublished PhD Thesis, Loughborough University, Department of Computer Science, 1996.

Rigas, D.I. & J.L. Alty. The use of music in a graphical interface for the visually impaired. In: *Proceedings of Interact '97, the International Conference on Human-Computer Interaction*, S. Howard, J. Hammond & G. Lindegaard (Eds.), pp. 228-235, Sydney: Chapman and Hall, 1997.

RNIB. *This is Moon*, RNIB. http://www.rnib.org.uk/braille/moonc.htm. 1996.

Sacks, A.H. & R. Steele. A journey from concept to commercialization - Lingraphica. *OnCenter Tech100 nology Transfer News* (5), (http://guide.stanford.edu/Publications/issue5.html#ling). 1993.

Schulz, B. & B. Wilhelm. Access to computerized line drawings with speech. In: *Computers for Handicapped Persons: Proceedings of the 3rd International Conference, ICCHP '92*, W.L. Zagler (Ed.), pp. 461-465, Vienna, Osterreichische Computer Gesellschaft. 1992.

Smith, S.L. & J.N. Mosier. *Design guidelines for user-system interface software*, Report ESD-TR-84-190, USAF Electronics Division, (http://www.info.fundp.ac.be/httpdocs/guidelines/ ??). 1984.

Steele, R.D. & M. Weinrich. Training of severely impaired aphasics on a computerized visual communication system. In: *Proceedings of Resna 8th Annual Conference*, pp. 320–322. 1986.

Stevens, R. *Principles for the design of auditory interfaces to present complex information to blind computer users*, DPhil Thesis, University of York, UK, 1996.

Strong, G.W. An evaluation of the PRC Touch Talker with Minspeak: Some lessons for speech prosthesis design. In: *Extra-Ordinary Human-Computer Interaction: Interfaces for Users with Disabilities*. A.D.N. Edwards (Ed.) New York: Cambridge University Press: pp. 47–57, 1995.

UN. The United Nations Declaration on the Rights of Disabled Persons. *Unesco Courier* 1: pp. 6–7, 1981.

Youngblut, C., R.E. Johnson, *et al. Review of Virtual Environment Interface Technology*, Report IDA Paper P-3186, Institute for Defense Analyses - IDA, (http://www.hitl.washington.edu/scivw/IDA/). 1996.

Zhang, J. A representational analysis of relational information displays. *International Journal of Human-Computer Studies* **45**: pp. 59–74, 1996.

7. AFFILIATION

A.D.N. Edwards, Department of Computer Science, University of York, York, UK, YO10 5DD

NIELS OLE BERNSEN

MULTIMODALITY IN LANGUAGE AND SPEECH SYSTEMS - FROM THEORY TO DESIGN SUPPORT TOOL

1. INTRODUCTION

This paper presents an approach towards achieving fundamental understanding of unimodal and multimodal output and input representations with the ultimate purpose of supporting the design of usable unimodal and multimodal human-human-system interaction (HHSI). The phrase 'human-human-system interaction' is preferred to the more common 'human-computer interaction' (HCI) because the former would appear to provide a better model of our interaction with systems in the future, involving (i) more than one user, (ii) a complex networked system rather than a (desktop) 'computer' which in most applications may soon be a thing of the past, and (iii) a system which increasingly behaves as an equal to the human users (Bernsen, 2000). Whereas the enabling technologies for multimodal representation and exchange of information are growing rapidly, there is a lack of theoretical understanding of how to get from the requirements specification of some application of innovative interactive technology to a selection of the input/output modalities for the application which will optimise the usability and naturalness of interaction. Modality Theory is being developed to address this, as it turns out, complex and thorny problem starting from what appears to be a simple and intuitively evident assumption. It is that, as long as we are in the dark with respect to the nature of the elementary, or unimodal, modalities of which multimodal presentations must be composed, we do not really understand what multimodality is. To achieve at least part of the understanding needed, it appears, the following objectives should be pursued, defining the research agenda of Modality Theory (Bernsen, 1993):

(1) To establish an exhaustive taxonomy and systematic analysis of the unimodal modalities which go into the creation of multimodal *output* representations of information for HHSI.

(2) To establish an exhaustive taxonomy and systematic analysis of the unimodal modalities which go into the creation of multimodal *input* representations of information for HHSI. Together with Step (1) above, this will provide sound foundations for describing and analysing any particular system for interactive representation and exchange of information.

B. Granström et al. (eds.), Multimodality in Language and Speech Systems, 93–148.

(3) To establish principles for how to legitimately combine different unimodal output modalities, input modalities, and input/output modalities for usable representation and exchange of information in HHSI.
(4) To develop a methodology for applying the results of Steps (1) – (3) above to the early design analysis of how to map from the requirements specification of some application to a usable selection of input/output modalities.
(5) To use results in building, possibly automated, practical interaction design support tools.

The research agenda of Modality Theory thus addresses the following general problem: given any particular set of information which needs to be exchanged between user and system during task performance in context, identify the input/output modalities which constitute an optimal solution to the representation and exchange of that information. As we shall see and as has become obvious from the literature on the subject through the 1990s, this is a hard problem, for two reasons. Firstly, already at the level of theory there are a considerable number of unimodal modalities to consider whose combinatorics, therefore, is quite staggering. Secondly, when it comes to applying the theory in development practice, the context of use of a particular application must be taken thoroughly into account in terms of task, intended user group(s), work environment, relevant performance and learning parameters, human cognitive properties, etc. A particular modality is not simply good or bad at representing a certain type of information – its aptness for a particular application very much depends on the context. This adds to the combinatorics generated by the theory an open-ended space of possibilities for consideration by the developer, a space which, furthermore, despite decades of HCI/HHSI research remains poorly mastered, primarily because such is the nature of engineering as opposed to abstract theory.

Given the many different and confusing ways in which the terms 'media' and 'modality' are being used in the literature, it should be made clear from the outset what these terms mean in Modality Theory.

A *medium* is the physical realisation of some presentation of information at the interface between human and system. Media are closely related to the classical psychological notion of the human "sensory modalities", i.e. vision, hearing, touch, smell, taste, and balance. Thus, the graphical medium is what humans or systems see, i.e. light, the acoustic medium is what humans or systems hear, i.e. sound, and the haptic medium is what humans or systems touch. Physically speaking, graphics comes close to being photon distributions, and acoustics comes close to being sound waves. In physical terms, haptics is obviously more complex than those two and no attempt will be made here to provide a physical description of haptics beyond stating that haptics involve touching. Media are symmetrical between human and system: a human hears (output) information expressed by a system in the acoustic medium, a system sees (input) information expressed by a human in the graphical medium (in front of a camera, for instance), etc. In the foreseeable future, information systems will mainly be using the three input/output media of graphics, acoustics and haptics. These are the media addressed by Modality Theory so far. To forestall a possible misunderstanding, the medium of graphics includes both text and "graphics" in the sense of images, diagrams, graphs etc. (see below).

The term *modality* (or *representational modality* as distinct from the sensory modalities of psychology) simply means "mode or way of exchanging information between humans or between humans and machines in some medium". The reason why any approach to multimodality is bound to need both of the notions of media and modalities is that media only provide a very coarse-grained way of distinguishing between the many importantly different physically realised kinds of information which can be exchanged between humans and machines. For instance, a graphical output image and a typed Unix output expression are both output graphics, or an alarm beep and a synthetic spoken language instruction are both output acoustics, even though those representations have very different properties which make them suited or unsuited, as the case may be, for different tasks, users, environments, etc. It seems obvious, therefore, that we need a much more fine-grained breakdown among available representational modalities than what is offered by the distinction between different media. The notion of representational modalities just introduced is probably quite close to that intended by many authors. As early as ten years ago, Hovy & Arens (1990), observed that, e.g., tables, beeps, written and spoken natural language may all be termed 'modalities' in some sense.

Some additional terms are clarified briefly to avoid misunderstandings later on. *Input* means interactive information going from A to B and which has to be decoded by B. A and B may be either humans or systems. Typically in what follows, A will be a human and B will be a system. It is thus taken for granted that we all know a lot about what can take place in an interaction in which both A and B are humans, or in which several humans interact together as well as interacting with a system. *Output* means interactive information going from B (typically the machine) to A (typically a human). The term *interactive* emphasises that A and B exchange information deliberately or that they communicate. In this central sense of 'interaction', it is *not* interaction when, e.g., a surveillance camera tracks and records an intruder unbeknownst to that intruder. It should also be noted that Modality Theory is about (representational) modalities and not about the *devices* which machines and humans use when they exchange information, such as hands, joysticks, or sensors. The positive implication is that the world of modalities is far more stable than the world of devices and hence much more fit for stable theoretical treatment. The negative implication is that Modality Theory in itself does not address the – sometimes tricky – issues of device selection which may arise once it has been decided to use a particular set of input/output modalities for an application to be built. On a related note, the theory has nothing to say about how to do the detailed design (aesthetically or otherwise) of *good* output presentations of information using particular modalities. As the colourful field of animated interface agents illustrates at present, it is one thing to safely assume that these virtual creatures have strong potential for certain kinds of application but quite another to demonstrate that potential through successful design solutions. Finally, it should be pointed out that when we refer to the issue of which modalities to use for exchanging information of some kind, 'information' means information in the abstract, as in 'medical data entry information', information in a new interactive game to be developed, or geographical information for the blind. Such descriptions are commonplace, and they leave more or less completely open the question of which modalities to use for the particular purpose at hand.

Modality Theory is, in fact, a century-old subject which easily antedates even the Babbage machine. People have interacted with information presentations on pyramids, in books or in magazines for a very long time. For instance, output modality analysis has a long tradition in the medium of (static) graphics. Outstanding examples are the results achieved on static graphic graphs (Bertin, 1983; Tufte, 1983, 1990). Given today's and tomorrow's input/output technologies, however, we need to address a much wider range of modalities and modality combinations. This is a truly collective endeavour. Modality Theory and the methodology for its practical application is an attempt to provide and illustrate a reasonably sound theoretical framework for integrating the thousands of existing and emerging individual contributions to our understanding of the proper use of modalities in interaction design and development.

This chapter addresses, at different levels of detail, all of the five points on the research agenda of Modality Theory described above, as follows. Section 2 presents the generation of the taxonomy for unimodal output modalities at several levels of abstraction. Section 3 proposes a draft standard representation format for modality analysis. Section 4 presents ongoing work on generating the taxonomy for input modalities. This part of the research agenda has proved to be hard and full of surprises. Section 5 presents our first full-scale application of the theory in its role as interaction design support. Finally, Section 6 concludes by discussing empirical and theoretical approaches for how to deal with the combinatorial explosion of modality combinations in multimodal systems. Due to space limitations, it has sometimes been necessary to refer to other publications for more detail.

For the obvious reason, the modality illustrations to be provided below are all presented in static graphics just like the present text itself. Current literature tends to focus on input/output modalities which are technically more difficult to produce, and which are less explored, than the static graphics modalities. It may be worthwhile to stress at this point, therefore, that all or most of the modality concept to be introduced below in fact do generalise to all possible modalities in the media of graphics, acoustics and haptics.

2. A TAXONOMY OF UNIMODAL OUTPUT

The taxonomy of unimodal output modalities to be presented is not the only one around although it appears to be the only one which has been generated from basic principles rather than being purely, or mainly, empirical in nature. In addition, its scope is as broad as that of any other attempt in the literature. A solid taxonomy based on decades of practical experience is Tufte's taxonomy of data graphics (Tufte, 1983). Twyman (1979) presents a taxonomy of static graphics representations (text, images, etc.). It is of wider scope than Tufte's taxonomy and, like the latter, based on long practical experience. Still in the static graphics domain, (Lohse et al., 1991) present a taxonomy which is based on experiments in which they studied how subjects intuitively classify sets of static graphic representations. Of much broader scope, comparable to that Modality Theory, are the lists of modalities and modality combinations in (Benoit et al., 2000). These lists simply enumerate modalities found in a large sample of the literature on multimodality from the 1990s.

A taxonomy of representational modalities is a way of carving up the space of forms of representation of information based on the observation that different modalities have different properties which make them suitable for exchanging different types of information between humans and systems. Let us assume that modalities can be either unimodal or multimodal and that multimodal modalities are combinations of unimodal modalities, i.e. can be completely and uniquely defined in terms of unimodal modalities. These assumptions suggest that if we want to adopt a principled approach to the understanding and analysis of multimodal representations, we have to start by generating and analysing unimodal representations. Generation comes first, of course. So the crucial issue at this point is how to generate the unimodal modalities. Basically, two approaches are possible, one purely empirical, the other hypothetico-deductive, i.e. through empirical testing of a systematic theory or hypothesis. Note that both approaches are empirical ones, just in different ways. Although the purely empirical approach has a strong potential for providing relevant insights and is being used widely in the field, it appears that no stable scientific taxonomy was ever created in a purely empirical fashion from the bottom up. If, for instance, experimental subjects are asked to spontaneously cluster a more or less randomly selected set of analogue static graphic representations (Lohse et al., 1991), the subjects may classify according to different criteria, they may be unable to express the criteria they use, and in the individual subject the criteria that are being applied may be incoherent. An alternative to the purely empirical approach is to generate modalities from basic principles and then test through intuition, analysis, and experiment whether the generated modalities satisfy a number of general requirements. If not, the generative principles will have to be revised. Let us adopt the generative approach in what follows. We want to identify a set of unimodal *output* modalities which satisfies the following requirements:

(1) *completeness,* such that any piece of, possibly multimodal, output information in the media of graphics, acoustics and haptics can be exhaustively described as consisting of one or more unimodal modalities;
(2) *uniqueness,* such that any piece of output information in those media can be characterised in only one way in terms of unimodal modalities;
(3) *relevance,* such that the set captures the important differences between, e.g., beeps and spoken language from the point of view of output information representation; and
(4) *intuitiveness,* such that interaction developers recognise the set as corresponding to their intuitive notions of the modalities they need or might need. Given the practical aims of Modality Theory, it is of crucial importance to operate with intuitively easily accessible notions without sacrificing systematicity.

To satisfy requirements (a) and (b) in particular, the generative process itself must be completely transparent. The four requirements differ in status as regards empirical testing of the generated taxonomy. Thus (d), on intuitiveness, is the more immediately accessible to evaluation. But even with respect to (d) as well as for (a) through (c), the theory can and should be exposed to more systematic empirical testing of various kinds.

The space of unimodal output representations can be carved up at different levels of abstraction. We have seen that already above, in fact, because the three media of graphics, acoustics and haptics may be viewed as a very general way of structuring the space of unimodal output representations. What will be proposed in the following is a downwards extensible, hierarchical generative taxonomy of unimodal output modalities which at present has four levels, a *super level*, a *generic level*, an *atomic level* and a *sub-atomic level*. In terms of the generative steps to be made, the generic level comes first. Thus, the taxonomy is based on a limited set of well-understood generic unimodal modalities. In their turn, the generic modalities are generated from sets of basic properties. An earlier version of the taxonomy to follow is (Bernsen 1994).

2.1. Basic Properties

We generate the generic-level unimodal output modalities from a small set of *basic properties* which serve to robustly distinguish modalities from one another within the taxonomy. The properties are: *linguistic/non-linguistic, analogue/non-analogue, arbitrary/non-arbitrary* and *static–dynamic*. In addition, distinction is made between the physical *media of expression* of graphics, acoustics, and haptics, each of which are characterised by very different sets of perceptual qualities (visual, auditory and tactile, respectively). These media determine the scope of the taxonomy. It follows that the taxonomy does not cover, for instance, olfactory and gustatory output representations of information which would appear less relevant to current interaction design. Thus, the scope of the taxonomy is defined according to the relevance requirement (c) above.

By taking those basic properties as points of departure, unimodal output modality generation starts from what are arguably the most general and robust distinctions among the capabilities of physically realised representations for representing information to humans. The set of basic properties has been chosen such that it is evident that their presence in, or absence from, a particular representation of information makes significant differences to the usability of that representation for interaction design purposes. For instance, the same linguistic message may be represented in either the graphical, acoustic, or haptic medium but the choice of medium strongly influences the suitability of the representation for a given design purpose and is therefore considered a choice between different modalities. So, the first justification for the choice of basic properties is their profoundly different capabilities for representing information. The second justification is that these basic properties appear to generate the right outcome, as we shall see, i.e. to eventually generate the unimodal output modalities which fit the intuitions and the relevance requirements which developers already have.

The basic properties may be briefly defined as follows, linguistic and analogue representations being defined in contrast to one another:

Linguistic representations are based on existing syntactic-semantic-pragmatic systems of meaning. Linguistic representations, such as speech and text, can, somehow, represent anything and one might therefore wonder why we need any other kind of modality for representing information in HHSI. The basic reason appears to be that linguistic representations lack the *specificity* which characterise analogue representations (Stenning & Oberlander, 1991; Bernsen, 1995). Instead, linguistic representations are *abstract* and *focused:* they focus, at some level of abstraction, on the subject matter to be communicated without providing its specifics. The cost of linguistic abstraction and focusing is to leave open an *interpretational scope* as to the nature of the specific properties of what is being represented. My neighbour, for instance, is a specific person who may have enough specific properties in the way he looks, sounds, and feels to distinguish him from any other person in the history of the universe, but you will not know much about these specifics from understanding the expression 'my neighbour'. The presence of abstract focus and the lack of specificity jointly generate the characteristic, limited expressive power of linguistic representations, whether these be static or dynamic, graphic, acoustic or haptic, or whether the linguistic signs used are themselves non-analogue as in the present text, or analogue as in iconographic sign systems such as hieroglyphs or Chinese. Linguistic representation therefore is, in an important sense, complementary to analogue representation. Many types of information can only with great difficulty, if at all, be rendered linguistically, such as how things, situations or events exactly look, sound, feel, smell, taste or unfold, whereas other types of information can hardly be rendered at all using analogue representations, such as abstract concepts, states of affairs and relationships, or the contents of non-descriptive speech acts. The complementarity between linguistic and analogue representation explains why their combination is so excellent for many representation purposes. For a detailed analysis of the implications of this complementarity for HHSI, see Bernsen (1995).

Analogue representations, such as images and diagrams, represent through aspects of similarity between the representation and what it represents. These aspects can be many, as in holograms, or few, as in a standard data graphics pie graph (or pie chart). Note that the sense of 'analogue' in Modality Theory is only remotely related to that of 'analogue (vs. digital)'. Being complementary to linguistic modalities, analogue representations (sometimes called 'iconic' or 'isomorphic' representations) have the virtue of specificity but lack abstract focus, whether they be static or dynamic, graphic, acoustic or haptic. Specificity and lack of focus, and, hence, lack of interpretational scope, generate the characteristic, limited expressive power of analogue representations. Thus, a photograph, haptic image, sound track, video or hologram representing my neighbour would provide the reader with large amounts of specific information about how he looks and sounds, which might only be conveyed linguistically with great difficulty, if at all. As already noted, the complementarity between linguistic and analogue representation explains why their (multimodal) combination is eminently suited for many representational purposes. Thus, one important use of language is to *annotate* analogue representations, such as a 2D graphic map or a haptic compositional diagram; and one important use of analogue representation is to *illustrate* linguistic text. In annotation, analogue representation provides the specificity which is being commented on in language; in illustration, language provides the generalities and abstractions which cannot be provided through analogue representation.

The distinction between *non-arbitrary* and *arbitrary representations* marks the difference between representations which, in order to perform their representational function, rely on an already existing system of meaning and representations which do not. In the latter case, the representation must be accompanied by appropriate representational conventions at the time of its introduction or else remain uninterpretable. Thus we stipulate things like "In this list, the boldfaced names are those who have already agreed to attend the meeting". In the case of non-arbitrary representations, such as when using the linguistic expressions of some natural language, introductory conventions are basically superfluous as the expressions already belong to an established system of meaning. It is not a problem for the taxonomy that representations, which used to be arbitrary, may gradually acquire common use and hence become non-arbitrary. Traffic signs may be a case in point.

Static representations and *dynamic representations* are mutually exclusive. However, the notion of static representation used in Modality Theory is not a purely physical one (what does not change or move relative to some frame of reference) nor is it a purely perceptual one (what does not appear to humans to change or move). Rather, static representations are such which offer the user *freedom of perceptual inspection*. This means that static representations may be decoded by users in any order desired and as long as desired. Dynamic representations are transient and do not afford freedom of perceptual inspection (Buxton 1983). According to this static/dynamic distinction, a representation is static even if it exhibits perceptible short-duration repetitive change. For instance, an acoustic alarm signal which sounds repeatedly until someone switches it off, or a graphic icon which keeps blinking until someone takes action to change its state, are considered static rather than dynamic representations. The implication is that some *acoustic* representations are static. A lengthy video that plays indefinitely, on the other hand, would still be considered dynamic because it does not exhibit short-duration repetitive change. The reason for adopting this not-purely-physical and not-purely-perceptual definition of static representation is that, from a usability point of view, and that is what interaction designers have to take into account when selecting modalities for their applications, the primary distinction is between representations which offer freedom of perceptual inspection and representations which do not. Just imagine, for instance, that your standard Windows GUI main screen were as dynamic as a lively animated cartoon. In that case, the freedom of perceptual inspection afforded by static graphics would be lost with disastrous results both for the decision-making process that precedes much interaction and for the interaction itself. This particular way of drawing the static-dynamic distinction does not imply, of course, that a blinking graphic image icon has exactly the same usability properties as a perceptually static one. Distinction between them is still needed and will have to be made internally to the treatment of static graphic modalities.

The *media* physically instantiate or embody representational modalities (se also Section 1). Through their respective physical instantiations, the various media are accessible through different sensory modalities, the graphic medium visually, the acoustic medium auditorily and the haptic medium tactilely. Different media have very different physical properties and are able to render very different sets of perceptual qualities. An important point which is sometimes ignored is that *all* of the perceptible physical properties characteristic of a particular medium, their respective scope of variation, and their relative cognitive impact are at our disposal when we use a certain representational modality in that medium. Standard typed natural language text, for instance, being graphical, can be manipulated graphically (boldfaced, italicised, coloured, rotated, highlighted, re-sized, textured, re-shaped, projected, zoomed-in-on etc.), and such manipulations can be used to carry information in context. Exactly the same holds for graphical images and other analogue graphical representations. This example shows that one should be careful when, or, indeed, preferably avoid, contrasting "text and graphics", because in the example just provided, the text *is* being graphically expressed. Text, or language more generally, need not be expressed graphically, however, but can be expressed acoustically (when read aloud) and haptically as well. Similarly, the reason why spoken language is so rich in information is that it exploits to the full the perceptible physical properties of the acoustic medium. We call these perceptible properties *information channels* and will return to them later (Section 3).

2.2. *Generating the Generic Level*

Given the basic properties presented in the previous section, the generation of the generic level of the taxonomy is purely mechanical, producing 48 (2x2x2x2x3) basic property combinations each of which represents a generic-level unimodal modality (Table 1). Each of the 48 generic unimodal modalities is completely, uniquely and transparently defined in terms of a particular combination of basic properties. Table 1 uses abbreviations to represent the basic properties. The meaning of these abbreviations should be immediately apparent. The term 'generic' indicates that unimodal modalities, as characterised at the generic level, are still too general to be used as a collection of unimodal modalities in an interaction designer's toolbox. The reason is that a number of important distinctions among different unimodal modalities cannot yet be made at the generic level (see Section 2.3).

All 48 unimodal modalities are perfectly possible forms of information representation. 48 unimodal output modalities at the generic level is a lot, especially since we are going to generate an even larger number when generating the atomic level of the taxonomy. However, closer analysis shows that it is possible to significantly reduce the number of generic modalities. The reductions to be performed are of two kinds. Both reductions are made with reference to the requirement of (current) relevance above. The first reduction is removal of modalities the use of which for interaction design purposes is in conflict with the purpose of Modality Theory. By

Table 1. The full set of 48 combinations of basic properties constituting possible unimodal output modalities at the generic level of the taxonomy. All modalities are possible ways of representing information.

	li	*-li*	*an*	*-an*	*ar*	*-ar*	*sta*	*dyn*	*gra*	*aco*	*hap*
1	x		x		x		x		x		
2	x		x		x		x			x	
3	x		x		x		x				x
4	x		x		x			x	x		
5	x		x		x			x		x	
6	x		x		x			x			x
7	x		x			x	x		x		
8	x		x			x	x			x	
9	x		x			x	x				x
10	x		x			x		x	x		
11	x		x			x		x		x	
12	x		x			x		x			x
13	x			x	x		x		x		
14	x			x	x		x			x	
15	x			x	x		x				x
16	x			x	x			x	x		
17	x			x	x			x		x	
18	x			x	x			x			x
19	x			x		x	x		x		
20	x			x		x	x			x	
21	x			x		x	x				x
22	x			x		x		x	x		
23	x			x		x		x		x	
24	x			x		x		x			x
25		x	x		x		x		x		
26		x	x		x		x			x	
27		x	x		x		x				x
28		x	x		x			x	x		
29		x	x		x			x		x	
30		x	x		x			x			x
31		x	x			x	x		x		
32		x	x			x	x			x	
33		x	x			x	x				x
34		x	x			x		x	x		
35		x	x			x		x		x	
36		x	x			x		x			x
37		x		x	x		x		x		
38		x		x	x		x			x	
39		x		x	x		x				x
40		x		x	x			x	x		
41		x		x	x			x		x	
42		x		x	x			x			x
43		x		x		x	x		x		
44		x		x		x	x			x	
45		x		x		x	x				x
46		x		x		x		x	x		
47		x		x		x		x		x	
48		x		x		x		x			x
	li	*-li*	*an*	*-an*	*ar*	*-ar*	*sta*	*dyn*	*gra*	*aco*	*hap*

contrast with the first reduction, the second reduction is not a removal of modalities but merely a fusion of some of them into larger categories. Both reductions are completely reversible, of course, simply by reinstating modalities from Table 1 which have been removed, or by re-separating modalities which were subjected to fusion. The reductions will be described in the following.

Table 2. 30 generic unimodal modalities result from removing from Table 1 the arbitrary use of non-arbitrary modalities of representation. The left-hand column shows the super level of the taxonomy. Modality theory notation has been added in the right-hand column.

SUPER LEVEL	*GENERIC UNIMODAL LEVEL*	*NOTATION*
I. Linguistic modalities	1. Static analogue sign graphic language	<li,an,-ar,sta,gra>
	2. Static analogue sign acoustic language	<li,an,-ar,sta,aco>
	3. Static analogue sign haptic language	<li,an,-ar,sta,hap>
<li,-an,-ar>	4. Dynamic analogue sign graphic language	<li,an,-ar,dyn,gra>
	5. Dynamic analogue sign acoustic language	<li,an,-ar,dyn,aco>
	6. Dynamic analogue sign haptic language	<li,an,-ar,dyn,hap>
	7. Static non-analogue sign graphic language	<li,-an,-ar,sta,gra>
	8. Static non-analogue sign acoustic language	<li,-an,-ar,sta,aco>
	9. Static non-analogue sign haptic language	<li,-an,-ar,sta,hap>
	10. Dynamic non-analogue sign graphic language	<li,-an,-ar,dyn,gra>
	11. Dynamic non-analogue sign acoustic language	<li,-an,-ar,dyn,aco>
	12. Dynamic non-analogue sign haptic language	<li,-an,-ar,dyn,hap>
II. Analogue modalities	13. Static analogue graphics	<-li,an,-ar,sta,gra>
	14. Static analogue acoustics	<-li,an,-ar,sta,aco>
	15. Static analogue haptics	<-li,an,-ar,sta,hap>
<-li,an,-ar>	16. Dynamic analogue graphics	<-li,an,-ar,dyn,gra>
	17. Dynamic analogue acoustics	<-li,an,-ar,dyn,aco>
	18. Dynamic analogue haptics	<-li,an,-ar,dyn,hap>
III. Arbitrary modalities	19. Arbitrary static graphics	<-li,-an,ar,sta,gra>
	20. Arbitrary static acoustics	<-li,-an,ar,sta,aco>
	21. Arbitrary static haptics	<-li,-an,ar,sta,hap>
<-li,-an,ar>	22. Dynamic arbitrary graphics	<-li,-an,ar,dyn,gra>
	23. Dynamic arbitrary acoustics	<-li,-an,ar,dyn,aco>
	24. Dynamic arbitrary haptics	<-li,-an,ar,dyn,hap>
IV. Explicit modality structures	25. Static graphic structures	<-li,-an,-ar,sta,gra>
	26. Static acoustic structures	<-li,-an,-ar,sta,aco>
	27. Static haptic structures	<-li,-an,-ar,sta,hap>
	28. Dynamic graphic structures	<-li,-an,-ar,dyn,gra>
<-li,-an,-ar>	29. Dynamic acoustic structures	<-li,-an,-ar,dyn,aco>
	30. Dynamic haptic structures	<-li,-an,-ar,dyn,hap>
SUPER LEVEL	*GENERIC UNIMODAL LEVEL*	*NOTATION*

Some modalities in Table 1 are inconsistent with the *purpose* of the taxonomy. Modality Theory in general and the taxonomy of unimodal output modalities in particular, serve the clear and efficient presentation and exchange of information. Given this purpose, the *arbitrary use of non-arbitrary representations* constitutes a capital sin in the context of interaction design. What this involves is providing a representation which already has an established meaning, with an entirely different meaning. For instance, arbitrary use of established linguistic expressions in a static graphic interface (Modality 13 in Table 1) should not occur in information systems output. To do so would be like wanting to achieve clear and efficient communication by letting 'yes' mean 'no' and vice versa. The result, as we know from children's games, is massive production of communication error and ultimate communication failure. Or if, for instance, a graphic interface designer lets iconic images of apples refer to ships on a graphic screen ocean map rather than using iconic images of ships for this purpose (assuming, among other things, that the ships do not carry apples), we have another case of using non-arbitrary representations arbitrarily. We also have a case of bad (i.e. confusing) interface design. This style of information representation is certainly meaningful and sometimes even useful, as in classical cryptography which makes use of the expressive strength of particular tokens belonging to some representational modality in order to mislead. Modality Theory, on the other hand, aims to support designers in making the best use of representational modalities for the purpose of clear and efficient presentation and exchange of information, through building on the expressive strengths of each. The taxonomy, therefore, simply does not address cryptography. What about passwords? It would seem that, in general, passwords are not cryptographic representations. They are just meant to be kept secret, and that is something else. Similarly, it is perfectly acceptable to use numbers arbitrarily in the sense of, for instance, arbitrarily assigning different numbers to players on a team. This is not in conflict with the meaning of numbers. Problems only start to arise if, say, a player is being assigned the number 3 and everybody is being told that the player has, in fact, number 2.

We thus have to remove columns 1-6, 13-18 and 25-30 from the Table 1 matrix. The remaining 30 unimodal output modalities are presented in a more explicit form in Table 2 which names each modality and shows its notation. Table 2 also shows the super level of the taxonomy (see below).

The second reductive step in generating the generic level of the taxonomy is *purely pragmatic* or practical rather than theoretical. The reduction of the number of unimodal modalities from 30 to 20 (Table 3) has been done uniquely in order to simplify the work involved in using the taxonomy for practical purposes, cf. the intuitiveness and relevance requirements above. The resulting taxonomy becomes less scholastic, as it were, and more usable. Table 3 integrates the presentation and analysis of *static* acoustic modalities with the presentation and analysis of *dynamic* acoustic modalities, and integrates the presentation and analysis of *static* haptic modalities with the presentation and analysis of *dynamic* haptic modalities. No modality information is lost in the process, so the completeness requirement is not being violated.

The practical reasons are as follows. Static acoustics, such as acoustic alarm signals, constitute a relatively small and reasonably well-circumscribed fraction of acoustic representations in whatever acoustic modality. For practical purposes, the presentation and analysis of the static acoustic modalities may without loss of information be integrated with that of the corresponding dynamic acoustic modalities which constitute the main class of acoustic representations. Similarly, dynamic haptics, such as the invention of dynamic Braille text devices where users do not have to move their fingers because the device pad itself changes dynamically to display new signs, currently constitute a relatively small fraction of haptic representations in whatever haptic modality. The dynamic haptics fraction may not be well circumscribed, however, and may be expected to grow dramatically with the growth of haptic output technologies. When this happens, we may simply re-instate the static/dynamic distinction in the haptic modalities part of the taxonomy.

Table 3. The 20 generic unimodal modalities resulting from pragmatic fusion of the static and dynamic acoustic modalities and the static and dynamic haptic modalities in Table 2.

SUPER LEVEL	*GENERIC UNIMODAL LEVEL*	*NOTATION*
I. Linguistic	1. Static analogue sign graphic language	<li,an,-ar,sta,gra>
modalities	2. Static analogue sign acoustic language Dynamic analogue sign acoustic language	<li,an,-ar,sta/dyn,aco>
	3. Static analogue sign haptic language Dynamic analogue sign haptic language	<li,an,-ar,sta/dyn,hap>
<li,-an,-ar>	4. Dynamic analogue sign graphic language	<li,an,-ar,dyn,gra>
	5. Static non-analogue sign graphic language	<li,-an,-ar,sta,gra>
	6. Static non-analogue sign acoustic language Dynamic non-analogue sign acoustic language	<li,-an,-ar,sta/dyn,aco>
	7. Static non-analogue sign haptic language Dynamic non-analogue sign haptic language	<li,-an,-ar,sta/dyn,hap>
	8. Dynamic non-analogue sign graphic language	<li,-an,-ar,dyn,gra>
II. Analogue	9. Static analogue graphics	<-li,an,-ar,sta,gra>
modalities	10. Static analogue acoustics Dynamic analogue acoustics	<-li,an,-ar,sta/dyn,aco>
<-li,an,-ar>	11. Static analogue haptics Dynamic analogue haptics	<-li,an,-ar,sta/dyn,hap>
	12. Dynamic analogue graphics	<-li,an,-ar,dyn,gra>
III. Arbitrary	13. Arbitrary static graphics	<-li,-an,ar,sta,gra>
modalities	14. Arbitrary static acoustics Dynamic arbitrary acoustics	<-li,-an,ar,sta/dyn,aco>
<-li,-an,ar>	15. Arbitrary static haptics Dynamic arbitrary haptics	<-li,-an,ar,sta/dyn,hap>
	16. Dynamic arbitrary graphics	<-li,-an,ar,dyn,gra>
IV. Explicit	17. Static graphic structures	<-li,-an,-ar,sta,gra>
modality structures	18. Static acoustic structures Dynamic acoustic structures	<-li,-an,-ar,sta/dyn,aco>
	19. Static haptic structures Dynamic haptic structures	<-li,-an,-ar,sta/dyn,hap>
<-li,-an,-ar>	20. Dynamic graphic structures	<-li,-an,-ar,dyn,gra>
SUPER LEVEL	*GENERIC UNIMODAL LEVEL*	*NOTATION*

Overall, at the generic and atomic levels combined, the proposed fusions reduce the number of modalities in the taxonomy by some 30 modalities. In a designer's toolbox, we want no more tools than we really need.

The 30 and 20 generic unimodal modalities of Tables 2 and 3, respectively, have been divided into four different classes at the super level, i.e. the linguistic, the analogue, the arbitrary, and the explicit structure modalities. The *super level* merely represents one convenient way of classifying the generic-level modalities among others, although, once laid down, it determines the overall surface architecture of the taxonomy. Other, equally useful, classifications are possible and can be used freely in modality analysis and taxonomy use, such as classifications according to medium or according to the static/dynamic distinction. Furthermore, the chosen super level modalities may not be deemed to be modalities proper (yet) as they lack physical realisation. This is a purely terminological issue, however. A point of greater significance is that, at the generic level, four of the linguistic modalities (Modalities 1 through 4 in Table 3) use analogue *signs* and four use non-analogue signs (Modalities 5 through 8 in Table 3). Basically, however, these are all primarily *linguistic,* and hence *non-analogue* representations because the integration of analogue signs into a syntactic-semantic-pragmatic system of meaning subjects the signs to sets of rules which make them far surpass the analogue signs themselves in expressive power. This may be the reason why all known, non-extinct iconographic languages have seen their stock of analogue signs decay to the point where it became difficult for native users to decode their analogue meanings.

2.3. Generating the Atomic Level

It may not be immediately obvious from Table 3 why the generic-level taxonomy cannot be used as a designer's toolbox of unimodal output modalities and does not meet the requirements of relevance and intuitiveness. This is partly due to the fact that some modalities are largely obsolete and hence irrelevant, such as the hieroglyphs subsumed by modality 1 in Table 3. Much more important, however, is the lack of intuitiveness of several of the modalities, such as Modality 9, 'static analogue graphics'. The lack of intuitiveness is due to the relatively high level of abstraction at which modalities are being characterised at the generic level. At the generic level, for instance, analogue static graphic *images* cannot be distinguished from analogue static graphic *graphs,* because both are subsumed by 'static analogue graphics'. In interaction design, however, these two modalities are being used for very different purposes of information representation and exchange. In another example, static graphic written *text* is useful for rather different purposes than is static graphic written *notation.* Yet both are subsumed by 'static non-analogue sign graphic language' at the generic level. Since the generic level does not make explicit such important distinctions among modalities, it is even difficult to put the completeness of the taxonomy to the test.

To achieve the relevance and intuitiveness required, which in their turn are preconditions for testing the completeness of the taxonomy, we need to descend at least one level in the abstraction hierarchy defined by the taxonomy. This is done by adding further distinctions among basic properties, thereby generating the atomic level of the taxonomy as presented in the static graphic conceptual diagram in Figure 1. In Figure 1, many of the generic modalities have several (equally unimodal) atomic modalities subsumed under them which inherit their basic properties and have distinctive properties of their own.

The new basic properties and distinctions to be introduced in order to generate the atomic level are specific to the super and generic level fragments of the taxonomy to which they belong. Where do these properties and distinctions come from? The generation of the atomic level follows the same principles as that of the

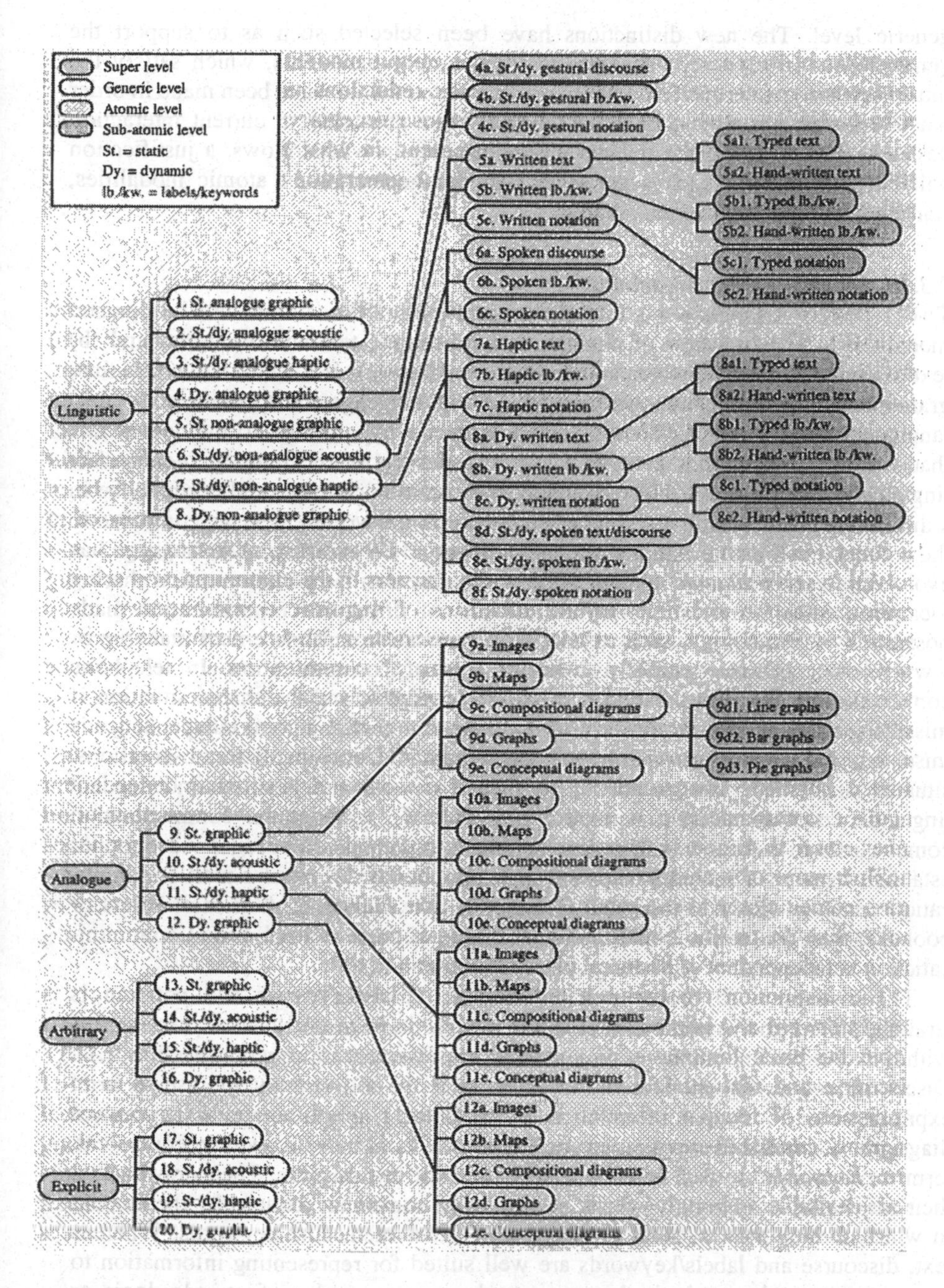

Figure 1. The taxonomy of unimodal output modalities. The four levels are, from left to right: super level, generic level, atomic level and sub-atomic level.

generic level. The new distinctions have been selected such as to support the generation of importantly different unimodal output modalities, which satisfy the intuitiveness requirement. In addition, pragmatic reductions have been made in order not to proliferate atomic modalities beyond those necessary in current interaction design, thus addressing the relevance requirement. In what follows, a justification will be presented for each super level segment generation of atomic modalities, starting with the linguistic modalities.

2.3.1. Linguistic Atomic Modalities

Two *types* of distinction go into the generation of the atomic level linguistic modalities. The first type of distinction is between (a) text and discourse, and (b) text/discourse, labels/keywords and notation. As to (a), it is a well-known fact that, grammatically and in many other respects, written language and spontaneous spoken language behave rather differently. This is due, we hypothesise, to the deeper fact that written language has evolved to serve the purpose of *situation independent* linguistic communication. The recipient of the communication would normally be in a different place, situation and time when decoding the written message compared to the context in which the author wrote the message. By contrast, spoken language has evolved to serve *situated* communication, the partners in the communication sharing location, situation and time. Hybrid situations of linguistic communication made possible by technology, such as telephone conversation, on-line e-mail dialogue or www chat, generate partially awkward forms of communication. In telephone conversation, the shared location is missing completely and the shared situation is missing more or less. In on-line e-mail dialogue and chat, temporal independence is missing and some situation sharing may be present. Generalising these observations, situated linguistic communication is termed *discourse* and situation independent linguistic communication is termed *text* (Table 4). Videophone communication comes closer to discourse than does telephone communication because videophones establish more of a shared situation than telephones do. Normal e-mail communication comes closer to the original forms of text exchange, such as mail letters or books, than do on-line e-mail and chat dialogue because normal e-mail communication is independent of partners' place, situation and time.

The distinction (b) between text/discourse, labels/keywords and notation is straightforward and important. *Text* and *discourse* have unrestricted expressiveness within the basic limitations of linguistic expressiveness in general (Section 2.1). Discourse and text modalities, however, tend to be too lengthy for use in brief expressions of focused information in menu lines, graph annotations, conceptual diagrams, command expressions etc. across media. *Labels* or, in another, equivalent, term, *keywords* are well suited and widely used for this purpose. Their drawback is their inevitable ambiguity which, at best, may be somewhat reduced by the context in which they appear, such as the context of other menu-line keywords. Whereas text, discourse and labels/keywords are well suited for representing information to any user who understands the language used, *notation,* such as first-order logic or xml, is for specialist users and always suffers from limited expressiveness compared to text and discourse. Text, discourse, labels/keywords and notation thus have importantly different but well-defined roles in interaction design across media and static-dynamic modalities.

Table 4. The atomic level unimodal linguistic modalities are shown in boldface in the right-hand column.

GENERIC UNIMODAL LEVEL	ATOMIC UNIMODAL LEVEL
1. Static analogue sign graphic language	Static gesture included in 4 a-c. Static text, labels/keywords, notation included in 5 a-c.
2. Static analogue sign acoustic language **Dynamic analogue sign acoustic language**	Included in 6 a-c.
3. Static analogue sign haptic language **Dynamic analogue sign haptic language**	Included in 7 a-c.
4. Dynamic analogue sign graphic language	Dynamic text, labels/keywords, notation included in 8 a-c. **4a. Static/dynamic gestural discourse** **4b. Static/dynamic gestural labels/keywords** **4c. Static/dynamic gestural notation**
5. Static non-analogue sign graphic language	**Static graphic spoken text, discourse, labels/ keywords, notation included in 8d-f.** **5a. Static graphic written text** **5b. Static graphic written labels/keywords** **5c. Static graphic written notation**
6. Static non-analogue sign acoustic language **Dynamic non-analogue sign acoustic language**	**6a. Static/dynamic spoken discourse** **6b. Static/dynamic spoken labels/keywords** **6c. Static/dynamic spoken notation**
7. Static non-analogue sign haptic language **Dynamic non-analogue sign haptic language**	**7a. Static/dynamic haptic text** **7b. Static/dynamic haptic labels/keywords** **7c. Static/dynamic haptic notation**
8. Dynamic non-analogue sign graphic language	**8a. Dynamic graphic written text** **8b. Dynamic graphic written labels/keywords** **8c. Dynamic graphic written notation** **8d. Static/dynamic graphic spoken text or discourse** **8e. Static/dynamic graphic spoken labels/keywords** **8f. Static/dynamic graphic spoken notation**

The second type of distinction involved in generating the atomic level is empirical in some restricted sense. That is, once the above distinctions have been made, it becomes an empirical matter to determine which important types of atomic linguistic modalities there are. This implies the possibility that Modality Theory might so far be missing some important type(s) of linguistic communication. However, testing made so far suggests that Table 4 presents all the important ones. In fact, the search restrictions imposed by the taxonomy does seem to enable close-to-exhaustive search in this case. When output by current machines, *gestural language* (4a-4c) is (mostly) dynamic and always graphic (even if done by a gesturing robot). Static gestural language is included in 4a-4c (see below). 5a-5c cover the original form of textual language, i.e. *static graphic written language.* The distinction between typed and hand-written static graphic written language belongs to the sub-atomic level (see below). 6a-6c cover the original form of discourse, i.e. *spoken language.* 7a-7c includes static and dynamic haptic language, such as Braille. The atomic modalities in Section 8 of Table 4 illustrate the empirical nature of atomic level generation. One might have thought that dynamic (non-analogue sign) graphic language simply includes 8a-8c, i.e. the dynamic versions of 5a-5c, such as scrolling text. However, Section 8 also includes graphically represented (non-acoustic) spoken language as produced, for instance, by a talking head or face, including read-aloud text and spoken discourse, labels/keywords and notation (8d-8f).

The pragmatic reductions of the linguistic atomic modalities are straightforward. As argued in Section 2.2, the fact that some written language uses analogue signs is ultimately insignificant compared to the fact that written language is a syntactic-semantic-pragmatic system of meaning. Written hieroglyphs and other iconographic textual languages, such as Chinese, and whether these are static or dynamic, graphic or haptic (Sections 1, 3 and 4 of Table 4), may therefore be fusioned with their non-analogue, non-iconographic counterparts without effects on interaction design. (The "glyphs" which have been invented for expressing multi-dimensional data points in graph space are rather forms of static graphic arbitrary modalities (Joslyn et al., 1995, Section 2.3.3)). Analogue speech sounds (onomatopoietica and others), by contrast (Section 2 of Table 4), constitute a genuine sub-class of speech. As such, they have been pragmatically included in Section 6 of Table 4. Static gestural language, such as the 'V' sign and many others, (Section 1 of Table 4), has been fusioned with dynamic gestural language (Section 4). Finally, the static graphic spoken language atomic modalities, such as a "frozen" talking head (Section 5), have been fusioned with their dynamic counterparts (Section 8). The result of this comprehensive set of fusions is shown as six triplets of atomic linguistic modalities in Figure 1. These modalities are shown in boldface in the right-hand column of Table 4. The strong claim of Modality Theory is that these modalities are all that interaction designers need in order to have a complete, unique, relevant and intuitive set of unimodal linguistic output modalities at the atomic level of abstraction. If more linguistic modalities are needed, they must either be generated from those at the atomic level and hence belong to some lower level of abstraction, such as the sub-atomic level, or they will re-appear by backtracking on some reduction (fusion) performed to obtain the modalities which presently constitute the linguistic atomic level.

The next section (2.3.2) will discuss *prototypical structure* and *continuity of representation,* phenomena which are prominent in the analysis of analogue representations and which need to be understood in order to avoid confusion in handling borderline issues of demarcation among different modalities. It should be noted that these phenomena are also present in the linguistic domain, so that, for instance, the issue over whether a certain representation is a collection of labels/keywords or is a notation may have to be decided by recourse to prototypical instances of labels/keywords and notation. In fact, prototypicality is a basic characteristic of conceptual structures. It follows that any theory of modalities will have to deal with the phenomenon.

2.3.2. Analogue Atomic Modalities

The analogue atomic modalities (Table 5) are generated without any pragmatic modality fusion or reduction. The generation is based on the concept of diagram and the distinction between (a) images, (b) maps, (c) compositional diagrams, (d) graphs and (e) conceptual diagrams. Diagrams subsume maps (b), compositional diagrams (c) and conceptual diagrams (e). The distinction between (a), (b), (c), (d) and (e) has been applied across the entire domain of analogue representation, whether static or dynamic, graphic, acoustic or haptic. What is needed, just like in the linguistic domain, is a justification of the distinctions, which have been introduced to generate the atomic level of analogue representation of information. Why are these distinctions the right ones with which to carve up the vast and complex space of analogue representation at the atomic level? For a start, it may probably be acknowledged that the concepts of images, maps, compositional diagrams, graphs and conceptual diagrams are both intuitively distinct and relevant to interaction design. At least two more questions need to be addressed, however. The first is whether the five concepts at issue exhaust the space of analogue atomic representation, cf. the completeness requirement in Section 2. The second question is how these concepts are defined so as to avoid overlaps and confusion when applying them to concrete instances in design practice, i.e. how distinct and mutually exclusive are these concepts in practice? Let us begin with the second question.

Table 5. The atomic level unimodal analogue modalities.

GENERIC UNIMODAL LEVEL	ATOMIC UNIMODAL LEVEL
9. Static analogue graphics	9a. Static graphic images 9b. Static graphic maps 9c. Static graphic compositional diagrams 9d. Static graphic graphs 9e. Static graphic conceptual diagrams
10. Static analogue acoustics Dynamic analogue acoustics	10a. Static/dynamic acoustic images 10b. Static/dynamic acoustic maps 10c. Static/dynamic acoustic compositional diagrams 10d. Static/dynamic acoustic graphs 10e. Static/dynamic acoustic conceptual diagrams
11. Static analogue haptics Dynamic analogue haptics	11a. Static/dynamic haptic images 11b. Static/dynamic haptic maps 11c. Static/dynamic haptic compositional diagrams 11d. Static/dynamic haptic graphs 11e. Static/dynamic haptic conceptual diagrams
12. Dynamic analogue graphics	12a. Dynamic graphic images 12b. Dynamic graphic maps 12c. Dynamic graphic compositional diagrams 12d. Dynamic graphic graphs 12e. Dynamic graphic conceptual diagrams

The exclusiveness (uniqueness) issue is particularly difficult in the analogue domain. The problem about exclusiveness in the analogue domain is that representations belonging to one modality, such as images (e.g. Figure 11), can often be manipulated to become as close as desired to representations belonging to several other modalities, such as compositional diagrams (Figure 2). Such *continuity of representation* is a well-known characteristic of many ordinary concepts and has been explored in prototype theory (Rosch, 1978). The point is that classical definitions using jointly necessary and sufficient conditions for specifying when an instance (or token) belongs to some category do not work well in the analogue domain. Instead, concept definitions have to rely on a combination of reference to *prototypical instances* (or paradigm cases) of a category combined with *characterising descriptions* that include pointers to *contrasts* between different categories. An important general implication is that the concepts of atomic and other modalities of Modality Theory cannot be fully intuitive in the sense of completely corresponding to our standard concepts. Any theory in the field would have to recognise this fact because such is the nature of the concepts, which we carry around in our heads. Ultimately, however, this is a desirable effect of theory. For instance, one of our present prototypical concepts of a static graphic image is the concept of a well-resembling 2D photograph of a person, landscape or otherwise (e.g. Figure 11). However, the static graphic images modality also includes 3D or 1D images, and these differ from those prototypes. In other words, Modality Theory can only meet the relevance and completeness requirements through some amount of *analytic generalisation,* which, in its turn, challenges the intuitiveness requirement somewhat. We shall see how the concept characterisations in the analogue domain work using abbreviated versions of the (unpublished) concept characterisations of Modality Theory which often run several pages per concept, excluding illustrations.

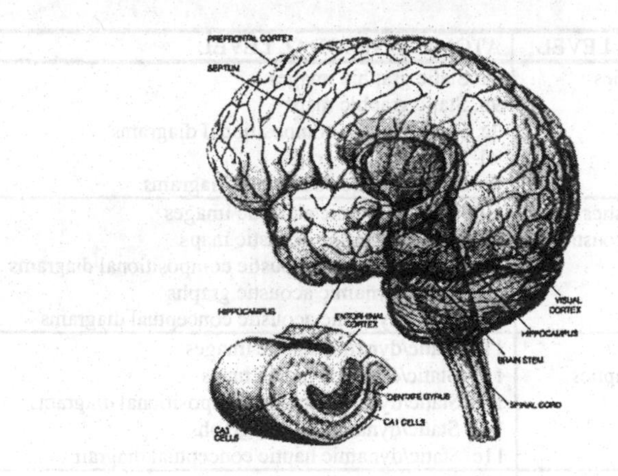

Figure 2. A prototypical compositional diagram: an annotated (hence bimodal) static 2D graphic representation of part of the structure of the brain.

A *diagram* may be briefly defined as an analytic analogue representation. A diagram provides an analytic account of its subject-matter, rather than an account of its mere appearance. This characterisation of diagrams will be expanded through characterisations of the various types of diagram below.

An *image* is an analogue representational modality which imitates or records the external form of real or virtual objects, processes and events by representing their physical appearance rather than serving analytical or decompositional purposes, such as those served by compositional diagrams. In the limit, as in ideal virtual reality output representations or standard input from real-life scenes, images allow realistic (quasi-) perception of the rich specific properties of objects, processes and events, which cannot easily be represented linguistically (Section 2.1). Images vary from high-dimensionality, maximally specific images to images whose specificity has been highly reduced for some purpose ('sketches'). Depending on the medium, images may represent non-perceivable objects, processes and events, whether these be too small, too big, too remote, too slow, too fast, beyond the human sensory repertoire (e.g. too high frequency, too low frequency), or normally hidden beneath some exterior, so that the objects, processes or events cannot themselves be perceived by humans. Images may also represent objects in a medium different from its 'normal' physical medium as when, for instance, acoustic information is being represented graphically (e.g. sonar images). Because images, considered on their own as unimodal representations, represent unfocused, association-rich 'stories', linguistic annotation is often needed to add focus and explanatory contents to the information they provide. In addition, many types of image, such as medical X-ray images, microscope images, or many types of sound pattern, require considerable skill for their interpretation. Figure 11 shows a prototypical image, i.e. a high-specificity 2D static graphic colour photograph of a person.

It may be observed from the above characterisation of images that images are being contrasted to their closest neighbour in analogue modality space, i.e. compositional diagrams (see below). Furthermore, it is pointed out in the image characterisation that images have limited value as stand-alone unimodal representations because of their lack of focus. For most interaction design purposes, images need linguistic annotation, which explains the intended point contributed by the image, so that the combined representation becomes bimodal. As a convenient, albeit coarse and relative generalisation which should be handled with care, unimodal modalities may be distinguished into *"independent" unimodal modalities* which can do substantial representational work on their own, and *"dependent" unimodal modalities* which need other modalities if they are to serve any, or most, representational purposes. Text and discourse modalities, for instance, are among the most independent unimodal modalities there are. Graphs, on the other hand, tend to be powerless in expressing information unless accompanied by other modalities. This issue will recur several times in what follows but a full discussion goes beyond the scope of this chapter. It still needs to be kept in mind that no unimodal modality has unlimited expressive power.

Compositional diagrams, such as an exploded representation of a wheelbarrow, are 'analytical images', i.e. they are analogue representations, which represent, using image elements, the structure or decomposition of objects, processes or events. The decomposition is typically linguistically labelled. Compositional diagrams focus on selective part-whole decomposition into, i.e., structure and function. Combinations of analogue representation and linguistic annotation in compositional diagrams vary from highly labelled diagrams containing rather abstract (i.e. reduced-specificity or schematic) analogue elements to highly image-like diagrams containing a modest amount of labelling. Highly labelled and abstract compositional diagrams, or compositional diagrams combining the representation of concrete and abstract subject matter, may occasionally be difficult to distinguish from conceptual diagrams (see below). To serve their analytic purpose, compositional diagrams typically involve important reductions of specificity, and they often use focusing mechanisms, saliency enhancement and dimensionality reduction (Bernsen, 1995). These selectivity mechanisms are used in order to optimise the compositional diagram for representing certain types of information rather than others. Figure 2 shows a rather high-specificity but otherwise prototypical compositional diagram, i.e. an annotated static 2D graphic representation of the structure of the brain. Figure 2 is a bimodal representation consisting of text labels and image elements.

Even more than images, compositional diagrams depend on linguistic annotation to do their representational job. Note also how compositional diagrams are being contrasted with their closest neighbours in analogue representation space, i.e. images and conceptual diagrams.

Maps are, in fact, a species of compositional diagrams, which are defined by their domain of representation. Maps provide geometric information about real or virtual physical objects and focus on the relational structure of objects and events, in order to present location information about parts relative to one another and to the whole. A prototypical map is a reduced-scale, reduced-specificity 2D graphic representation of part of the surface of the Earth, showing selected, linguistically labelled features such as rivers, mountains, roads and cities, and designed to enable travellers to find the right route between geographical locations. Maps may otherwise represent spatial layout of any kind, being on occasion difficult to distinguish from images and (non-map) compositional diagrams. Figure 3 shows a non-prototypical map, i.e. a unimodal, highly specificity-reduced static 2D graphic representation of the Copenhagen subway system. Only the topology and the relative positions of lines and stations have been preserved. The unimodality of the map in Figure 3 makes it uninterpretable for all but those possessing quite specific background knowledge, which enables them to supply the information, which is missing in the representation.

Maps are thus a species of compositional diagrams, sharing most of the properties of these as described above. Maps have been included in the taxonomy because they are quite common and application-specific, and because of the robustness of the map concept. We seem to think in terms of maps rather than in terms of 'a-certain-sub-species-of-compositional diagram'. A taxonomy of unimodal analogue modalities that ignores this fact is likely to be less relevant and intuitive than a taxonomy, which respects the fact while preserving analytic transparency.

Graphs represent quantitative or qualitative information through the use of analogue means which typically bear no *recognisable* similarity to the subject matter or domain of the representation. The quantitative information is statistical information or numerical data which may either be gathered empirically or generated from theories, models or functions. Their analogue character makes graphs well suited for facilitating users' identification of global data properties through making comparisons, perceiving data profiles, spotting trends among the data, perceiving tempo

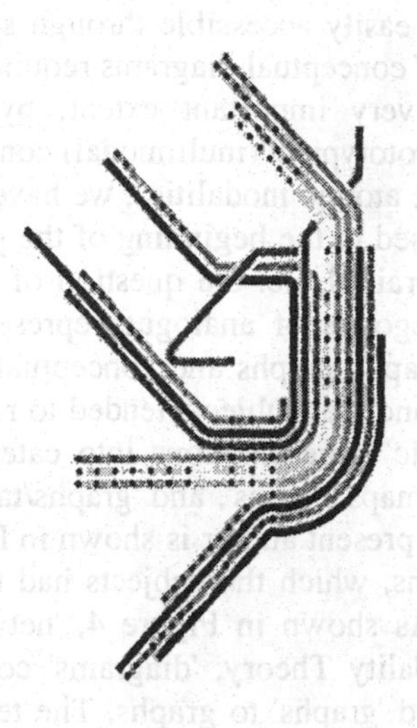

Figure 3. A non-prototypical map: a unimodal, highly specificity-reduced static 2D graphic representation of the subway system of Copenhagen.

ral developments in the data, and/or discovering new relationships among data, and hence supports the analysis of, and the reasoning about, data information. Whilst quantitative data can in principle be represented linguistically and are often presented in tables (see below), the focused and non-specific character of linguistic representation makes this form of representation ill suited to facilitate the interpretation of global data properties. Given their primarily abstract analogue nature, graphs virtually always require clear and detailed linguistic annotation, consistent with the analogue representation, for their interpretation. Thus, graphs are in practice (at least) bimodal modalities. Graphic graphs frequently incorporate graph space grids and other explicit structures (see below), which makes them trimodal modalities. The graph notion is quite robust and does not require contrasting with other analogue modalities - it has no close neighbours in analogue representation space. The huge diversity of graph representations requires a sub-atomic expansion of at least the static graphics graph node of the taxonomy (Section 2.5).

Conceptual diagrams use various analogue representational elements to represent the analytical decomposition of an abstract entity such as an organisation, a family, a theory or classification, or a conceptual structure or model. Thus, conceptual diagrams enhance the linguistic representation of abstract entities through analogue means, which facilitate the perception of structure and relationships. Conceptual diagrams constitute an abstract counterpart to compositional diagrams. The abstract, not primarily spatio-temporal representational purpose, and the decompositional purpose of conceptual diagrams jointly mean that conceptual diagrams require ample linguistic annotation and hence are at least bimodal. The role of the analogue elements in conceptual diagrams is to make the diagram's abstract subject-matter more easily accessible through spatial structure and layout. The abstract subject matter of conceptual diagrams requires that the information they represent be carried, to a very important extent, by the linguistic modalities involved. Figure 1 shows a prototypical (multimodal) conceptual diagram.

In presenting the analogue atomic modalities, we have so far concentrated on the question of exclusiveness raised in the beginning of the present section. Let us now address the second question raised, i.e. the question of completeness. The present taxonomy assumes four categories of analogue representation: images, compositional diagrams (including maps), graphs and conceptual diagrams. In an empirical study, Lohse et al. (1991) found that subjects tended to robustly categorise a variety of analogue 2D static graphic representations into categories, which they termed 'network charts', 'diagrams', 'maps', 'icons', and 'graphs/tables'. A free-hand drawing of their findings made by the present author is shown in Figure 4. It should be noted that the pile of representations, which the subjects had to classify, did not include any static graphic images. As shown in Figure 4, 'network charts' correspond to conceptual diagrams in Modality Theory, 'diagrams' correspond to compositional diagrams, 'maps' to maps, and 'graphs' to graphs. The terminology in the field has not been standardised, so there is nothing unexpected about these variations in terminology. As no images were presented to the subjects by Lohse et al. (1991), we can ignore images in what follows. Disregarding for the moment 'icons' and 'tables' which have not been discussed above, the correspondence between the result of Lohse et al. and the present taxonomy is very close indeed, at least in the domain of static graphics representation. However, Figure 4 also includes the well-known representation types 'icons' and 'tables', suggesting that the taxonomy above is not complete or exhaustive. So, what is the status of icons and tables in Modality Theory? The answer is that the theory provides what is arguably some much needed analytic generalisations concerning icons and tables.

Tables, although clearly distinct from any of the atomic modalities considered above, do not constitute a separate representational modality. Rather, tables are a convenient way of structuring information represented in most graphic or haptic modalities. Tables are a particular type of *modality structure* rather than a modality. Tables are often bimodal, as in prototypical 2D static graphic tables, which combine typed language with explicit structures (cf. Tables 1 through 5 above). Note, however, that explicit structures are not necessary constituents of tables. Simple tables can be elegantly presented without any use of explicit structures at all, just by appropriate spatial distribution of the tabular information. The fact that the subjects in Lohse et al. (1991) combined graphs and tables into one category is probably due to the fact that graph information can often be represented in tables, just like tabular information is often used to generate graphs. However, the fact that the abstract information contents of a table is sometimes equivalent to the information contents of a graph does not address the question of when it is preferable to represent that information in a table and when it is preferable to represent the information in a graph. Depending on the nature of the information and the purpose of the representation, graphs may be preferable to tables or vice versa (Tufte, 1983). Graphs and tables are, therefore, different forms of information representation. Moreover, tables may contain much else besides standard spreadsheet quantitative information, such as text, labels/keywords, notation, static or dynamic images, or graphs. The latter tables, such as a table showing a variety of dynamic graphics talking heads, do not have any graphs corresponding to them. In conclusion, tables can be uniquely defined in terms of unimodal modalities, i.e. as a particular, and quite general, way of structuring them, and tables are clearly distinct from graphs. *Lists* are another type of modality structure, which is different from, but related to, tables.

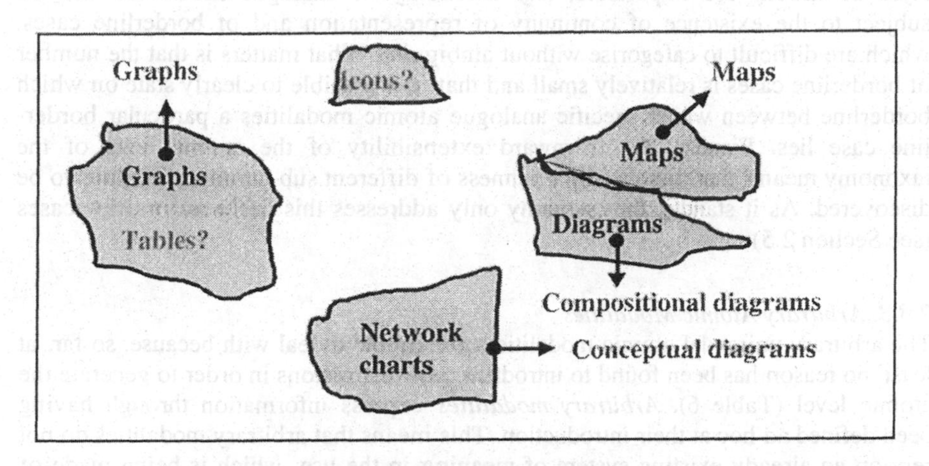

Figure 4. Subjects' classifications of analogue representations after Lohse et al. (1991). Arrows point to the corresponding terms used in Modality Theory. Question marks indicate phenomena, which are not modalities but something else.

Like lists and tables, *icons* do not constitute a separate modality. Rather, icons are "generalised labels/keywords" and the generalisation reaches far beyond 'icons' in the prototypical sense of static 2D graphic representations. Just like a label or keyword, an *icon* is a singular representation or expression, which normally has one intended meaning only, and which is subject to ambiguity of interpretation. Any token of any modality, it would appear, can be *used* as an icon, even a piece of text. Being an icon is, rather, a specific *modality role,* which can be assumed by any modality token. It would therefore be quite misleading to consider icons as a separate kind of modality. This means that icons are covered by the taxonomy to the extent that the taxonomy is exhaustive or complete. In other words, like tables and lists, icons can be uniquely defined in terms of unimodal modalities. Still, as Lohse found, icons are different from other modalities. The above analysis shows that the difference can be expressed in terms of the particular role, which a modality assumes when being used as an icon.

In conclusion, the correspondence between the taxonomy of analogue 2D static graphic modalities and the empirical results of Lohse et al. (1991) is now complete. Until an entirely different taxonomy of the space of analogue representation comes forward, and that has not happened yet, the present taxonomy would appear to be partially empirically confirmed. The results of Lohse et al. (1991) have confirmed the assumption that the four concepts of images, compositional diagrams (including maps), graphs, and conceptual diagrams exhaust the space of analogue atomic representation. It should still be kept in mind, however, that completeness does not imply exclusiveness. We have seen that classical-style definitions of analogue and other modalities are hardly possible. This implies that borderline cases will inevitably occur. For instance, is a static graphic image one part of which is labelled by a single typed label only, an image or a compositional diagram? But if classical-style definitions are impossible, *any* taxonomy of analogue modalities will be subject to the existence of continuity of representation and of borderline cases, which are difficult to categorise without ambiguity. What matters is that the number of borderline cases is relatively small and that it is possible to clearly state on which borderline between which specific analogue atomic modalities a particular borderline case lies. Finally, the downward extensibility of the atomic level of the taxonomy means that there is still a richness of different sub-atomic modalities to be discovered. As it stands, the taxonomy only addresses this richness in a few cases (see Section 2.5).

2.3.3. Arbitrary Atomic Modalities

The arbitrary unimodal atomic modalities are simple to deal with because, so far, at least, no reason has been found to introduce new distinctions in order to generate the atomic level (Table 6). *Arbitrary modalities* express information through having been defined ad hoc at their introduction. This means that arbitrary modalities do not rely on an already existing system of meaning in the use, which is being made of them. Arbitrary modalities are therefore by definition non-linguistic and non-analogue. As argued in Section 2.1, it is against the purpose of the taxonomy that non-arbitrary modalities be used arbitrarily. This imposes rather severe restrictions on which representations may be used arbitrarily. Nonetheless, arbitrary modalities can be very useful for representing information and we use them all the time. Information channels, in particular, are often useful for assuming arbitrary roles (Section 3). In general, any information channel in any medium can be arbitrarily assigned a specific meaning in context. This operation is widely used for expressing information in compositional diagrams, maps, graphs and conceptual diagrams and is illustrated in Figures 5 and 10 below. In another example, arbitrary modalities can be used to express acoustic alarms in cases where the only important point about the alarm is its relative saliency in context.

Table 6. The atomic level unimodal arbitrary modalities are identical to those at the generic level.

GENERIC UNIMODAL LEVEL	ATOMIC UNIMODAL LEVEL
13. Arbitrary static graphics	See generic level
14. Arbitrary static acoustics Dynamic arbitrary acoustics	See generic level
15. Arbitrary static haptics Dynamic arbitrary haptics	See generic level
16. Dynamic arbitrary graphics	See generic level

2.3.4. Explicit Structure Atomic Modalities

As in the case of the arbitrary atomic modalities, no reason has been found to introduce new distinctions in order to generate a larger set of explicit structure modalities at the atomic level than was already present at the generic level (Table 7). *Explicit structure modalities* express information in the limited but important sense of explicitly marking separations between modality tokens. Explicit structure modalities rely on an already existing system of meaning and are therefore non-arbitrary. This is because the purpose of explicit markings is immediately perceived. Explicit structure modalities are non-linguistic and non-analogue. Despite the modest amount of information conveyed by an explicit structure, these structures play important roles in interaction design. One such role is to mark distinction between different groupings of information in graphics and haptics. This role antedates the computer. Another, computer-related role is to mark functional differences between different parts of a graphic or haptic representation. Static graphic windows, for instance, are based on arbitrary structures, which inform the user about the different consequences of interacting with different parts of the screen at a certain point (Figure 12).

Table 7. The atomic level unimodal explicit structure modalities are identical to those at the generic level.

GENERIC UNIMODAL LEVEL	ATOMIC UNIMODAL LEVEL
17. Static graphic structures	See generic level
18. Static acoustic structures Dynamic acoustic structures	See generic level
19. Static haptic structures Dynamic haptic structures	See generic level
20. Dynamic graphic structures	See generic level

2.4. The Generative Power of the Taxonomy

The hypothesis, which has been confirmed up to this point in the development of Modality Theory and which is inherent to the atomic level of the taxonomy of unimodal output modalities, is a rather strong one. It is that the atomic level fulfils the requirements of completeness, uniqueness, relevance and intuitiveness stated above. Any multimodal output representation can be exhaustively characterised as consisting of a combination of atomic-level modalities.

Assuming that the atomic level of the taxonomy of unimodal output modalities has been generated successfully, an interesting implication follows. Space has not allowed the definition of each individual atomic modality presented in Tables 4 through 7 above What have been described are, rather, the principles that were applied in generating the atomic level and the new distinctions introduoed in the process. However, what has been generated goes far beyond the generative apparatus so far described. This is because the distinctions introduced in generating the atomic level get "multiplied" by the static/dynamic distinction and the distinction between different media of expression. The particular atomic modalities generated are the results of this multiplication process. Each atomic modality is distinct from any other and has a wealth of properties. Some of these are inherited from the modality's parent nodes at higher levels of abstraction in the taxonomy. Other properties specifically belong to the atomic modality itself and serve to distinguish it from its atomic-level neighbours. One way to briefly illustrate the generative power of the taxonomy is to focus on atomic modalities which are yet to become used in interaction design; which hold unexploited potential for useful information representation; which have not yet been discovered as representational modalities; or which are so "exotic" as to appear difficult to even exemplify for the time being. As several critics have noted, the very existence of just one such "exotic" unimodal modality goes against the requirement of relevance above. Before pragmatically removing them, however, which is easy as explained in the case of other modalities above, they should be scrutinised for potential relevance. In what follows, some of the generated unimodal output modalities, which fit the above descriptions are briefly commented upon.

Like any other atomic modality, *gestural notation* is a possible form of information representation. Except for use in brief messages, examples of gestural notation may be hard to come across. The reason probably is that notation, given its non-naturalness as compared to natural language, normally requires freedom of perceptual inspection to be properly decoded. Like *spoken language notation*, gestural notation would normally be dynamic and hence does not allow freedom of perceptual inspection. This leads to the prediction that, except for brief messages in dynamic notation, *static* gestural and spoken notation would be the more usable varieties (cf. Figure 9). For the same reasons, there would seem to be little purpose in using lengthy dynamic written notation, except for specialists capable of decoding such notation on-line, such as Morse-code specialists. Such specialists might find uses for lengthy gestural and spoken notation as well. If (acoustic) spoken notation is being expressed as synthetic speech, the specialists might need support from graphic spoken notation in order to properly decode the information expressed.

In the analogue atomic modalities domain, *acoustic images* are becoming popular, for instance in the 'earcon' modality role. *Acoustic graph-like images* have important potential for representing information in many domains other than, e.g., those of the clicking Geiger counter or the pinging sonar. The potential of *acoustic graphs* proper would seem to remain largely unexplored, except as redundant representations accompanying, e.g., static graphics bar graphs shown on TV. *Acoustic maps* appear to have some potential for representing spatial layout. *Acoustic compositional diagrams* offer interesting possibilities. Think, for instance, of a system for supporting the training of car repair trainees. Acoustic diagnosis plays an important role in the work of skilled car repairers. The training system might take apart the relevant diagnostic noises into their components, explain the causes of the component sounds and finally put these together again in training-and-test cycles. *Acoustic conceptual diagrams* may appear not to have any clear application potential. Yet it is possible to map, for instance, the different inheritance levels of a static graphic inheritance hierarchy into different keys. Primarily for reasons of technology and cost, output *dynamic analogue haptics* appears to be mostly unexplored territory, whether in the form of images, maps, compositional diagrams, graphs or conceptual diagrams. Yet their potential for special user populations would appear considerable. *Dynamic analogue graphics* are extremely familiar to us but still has great unused potential. Immersive virtual reality must combine dynamic, perceptually rich analogue graphics, acoustics and haptics.

Arbitrary static graphics, acoustics and *haptics* are widely used already. It is less obvious how much we shall need their dynamic counterparts in future applications. A ringing telephone, of course, produces arbitrary dynamic acoustics, and a vibrating mobile phone produces arbitrary dynamic haptics. Beyond such saliency-based applications, however, it is less clear which information representation purposes might be served by the dynamic arbitrary atomic modalities.

Finally, in the explicit structure domain, *static graphic explicit structures* are as commonplace as static graphics itself. *Dynamic graphic explicit structures* are in use as focusing mechanisms, for instance, for encircling linguistic or analogue graphic information of current interest during multimodal graphics and spoken language presentations. *Static* and *dynamic haptic explicit structures* have unexplored potential for the usual (technology and cost) reasons. In fact, it is only in the singular case of *acoustic explicit structures* that we have had problems coming up with valid examples. It is common, for instance, in spoken language dialogue applications to use beeps to indicate that the system is ready to listen to user input. However, as these beeps do not rely on an already existing system of meaning, they rather exemplify the use of arbitrary dynamic acoustics. Still, there might be useful functions out there for acoustic explicit structures.

It may be concluded that virtually all of the unimodal atomic output modalities in the taxonomy hold a claim to belong to a designer's toolbox of output modalities.

2.5. *Generating the Sub-Atomic Level of the Taxonomy*

Explicit completeness at any level of the taxonomy is still limited by level of abstraction and hence by the number of basic properties which has been introduced to generate that level. The generic level is as complete as the atomic level, but the former is less intuitive and relevant than the latter. One virtue of the taxonomy is its unlimited downward extensibility. That is, once the need has become apparent for distinguishing between different unimodal modalities subsumed by an already existing modality, further basic properties can be sought which might help generate the needed distinctions. In the domains of arbitrary and explicit structure modalities, this possibility remains unused already at the generic level (Sections 2.3.3 and 2.3.4). Below the atomic level, however, such as at the sub-atomic level of the taxonomy, there is still much representational diversity to be identified and used when required by theory and/or technology. In what follows, the purpose is merely to illustrate possibilities. The reader is invited to generate some of the other sub-atomic parts of the taxonomy, which already beckon to be generated. Table 8 shows how the principle of extensibility has been applied to static and dynamic graphic written text through the simple distinction between *typing* and *hand-writing* (cf. Figure 1). This extension may not be terribly important to output modality choice in interaction design except for reasons of aesthetic design, which go beyond Modality

Theory. The extension is important, however, to the developer's choice of *input* modalities.

Table 8. The sub-atomic level unimodal graphic written language modalities.

ATOMIC UNIMODAL LEVEL	SUB-ATOMIC UNIMODAL LEVEL
5a. Static graphic written text	5a1. Static graphic typed text 5a2. Static graphic hand-written text
5b. Static graphic written labels/keywords	5b1. Static graphic typed labels/keywords 5b2. Static graphic hand-written labels/keywords
5c. Static graphic written notation	5c1. Static graphic typed notation 5c2. Static graphic hand-written notation
8a. Dynamic graphic written text	8a1. Dynamic graphic typed text 8a2. Dynamic graphic hand-written text
8b. Dynamic graphic written labels/ keywords	8b1. Dynamic graphic typed labels/keywords 8b2. Dynamic graphic hand-written labels/keywords
8c. Dynamic graphic written notation	8c1. Dynamic graphic typed notation 8c2. Dynamic graphic hand-written notation

Table 9 shows what is still a hypothetical application of the principle of extensibility in the domain of static graphic graphs (cf. Figure 1). This example is much more complex than the one in Table 8. Static graphic graphs are extremely useful for representing quantitative information. The domain has been the subject of particularly intensive research for decades with the result that the atomic modality 'static graphic graphs' has become much too coarse-grained a notion for handling the large variety of information representations that exist. In fact, static graphic graph theory is one of the earliest examples of systematic work on a methodology for mapping from requirements specification to modality choice (e.g. Bertin, 1983; Tufte, 1983, 1990; Lockwood, 1969; Holmes, 1984). Even in this limited "corner" of the domain addressed by Modality Theory, however, there is still no consensus on taxonomy of static graphic graphs.

Table 9. The sub-atomic level unimodal static graphic graph modalities.

ATOMIC UNIMODAL LEVEL	SUB-ATOMIC UNIMODAL LEVEL
9d. Static graphic graphs	9d1. Line graphs 9d2. Bar graphs 9d3. Pie graphs

Our current hypothesis is that distinguishing between three basic types of graphs is sufficient for analysing the capabilities and limitations for information representation of all possible static graphic graphs. To avoid confusion it seems necessary to distinguish, in addition, between standard graphs and enhanced graphs. The *standard graphs* are: line graphs, bar graphs and pie graphs. This may appear at once both trivial and controversial. Trivial, because these graph types, as ordinarily understood, are the most common among all static graphic graph modalities. Controversial, both in so far as Tufte (1983) argues that nobody needs pie graphs and because different authors tend to provide different, and always considerably longer, lists of graph types. Typically, however, these lists are based on examples rather than careful definition. It would seem that the proposed graph types have not been sufficiently analysed as to their information representation capabilities. In addition, little attempt has been made to achieve reasonably generalised concepts of each type claimed to be basic to information representation. Finally, as is the norm rather than the exception in the modalities field, data graph terminology remains confusing and without any clear convergence towards a standard.

A standard *line graph* ('fever graph', 'curve chart') represents data points, whether empirical or generated, in a 1D, 2D or 2 1/2D/3D graph space, which is normally defined by co-ordinate axes. The term 'line graph' is somewhat misleading in referring to this class of graphs. The term derives from the fact that prototypical line graphs are 2D graphs in which the data points have been connected or computed over, the result being expressed in one or more lines which show how the changes in one quantity (dependent variable) are related to changes in another quantity (independent variable). However, discrete, non-continuous data point patterns need not be connected for a graph to be a line graph ('scatter diagrams/plots/graphs', 'dot charts'). In 1D no connecting lines are possible, whereas data point connections in 2 1/2D/3D are most often done using curved surfaces (Figure 5). Lines in 2D line graphs may be replaced by curved surfaces ('surface charts'). Cyclical data may be represented in circular line graphs using a circular co-ordinate system. Line graphs are good at representing large data sets with variability, showing flow, profiles, trends, history and projections, and are good for time-related data.

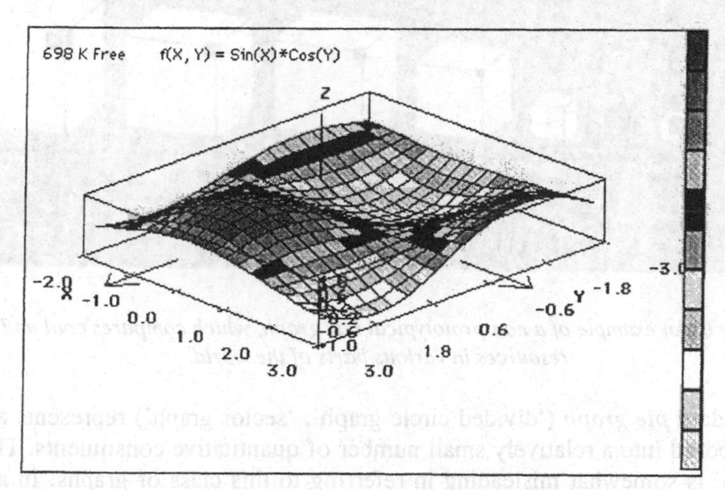

Figure 5. 2 1/2D static graphic surface function line graph using arbitrary colour coding for distinguishing hills and basins of the function.

A prototypical *bar graph* ('column graph') represents a small number of separate quantities lying within a comparative range in a 2D graph space. The term 'bar graph' is somewhat misleading in referring to this class of graphs. In a prototypical bar graph, horizontally or vertically aligned 2D bars are being used to represent quantities through their length. However, the bars may be replaced by other geometrical shapes in 1D, 2D, 2 1/2D or 3D whose length, area or volume represents the quantities in question (Figure 6). In 'circle graphs', for instance, it is not the length of the bar but the area of the circle, which represents the information. In 2D 'histograms' ('step charts'), the bars touch or have been replaced by stepwise curves and there is usually no spacing between the columns. Bars may be non-aligned ('float' or 'slide'), diagonal, radiate from the centre of a circle or from the circumference towards the centre, be stacked in a ('population') pyramid, shown at an angle or as receding towards the horizon, folded to encompass out-of-range quantities, etc. Bar graphs are good at enabling comparison, particularly when horizontally or vertically aligned, and the spotting of relationships among relatively small numbers of individual quantities. If the represented data are of a lower dimensionality than the information-carrying dimensions of the (generalised) 'bars' used to represent them, there is a high risk of generating misinformation. Similarly, humans are bad at correctly perceiving proportional relationships between areas or volumes. Work on 3D virtual reality graphs has identified interesting problems in using 3D graphic graphs, such as occlusion and perspective (Mullet & Schiano, 1995).

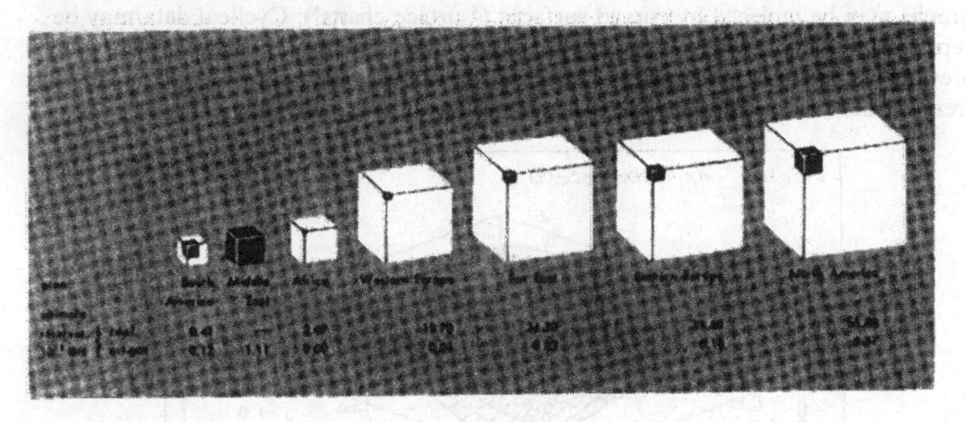

Figure 6. An example of a non-prototypical bar graph, which compares coal and oil resources in various parts of the world.

A standard *pie graph* ('divided circle graph', 'sector graph') represents a whole as decomposed into a relatively small number of quantitative constituents. The term 'pie graph' is somewhat misleading in referring to this class of graphs. In a prototypical pie graph, 2D or 2 1/2D circles or pies are being used to represent quantities through their percentage-wise partitioning. However, the circles or pies may be replaced by other regular geometrical shapes including, e.g., lines, squares, rectangles, triangles, ellipses, boxes, spheres, cylinders, cones, etc. Pie graphs are good at representing small numbers of parts of some whole with a view to comparing them. However, humans are bad at correctly perceiving proportional relationships between areas or volumes.

The above characterisations of sub-atomic-level static graphic graphs illustrate the need to refer to prototypes, the need to conceptually generalise quite strongly beyond these, the approach to defining graphs from the types of quantitative information which each type is best suited to represent, and the "reduction" of graph types proposed in the literature into a small number of graph modalities. If the proposed generation works, it can be rather straightforwardly generalised to include sub-atomic *dynamic* graphic graphs and static and dynamic *haptic* graphs. The latter sub-atomic generations are not shown in Figure 1. It is an interesting question whether all *acoustic* graphs must be line graph modalities or whether useful applications can be found for acoustic bar graphs and pie graphs.

Enhanced graphs are multimodal representations which not only, as in all interpretable graphs, in bimodal fashion combine an analogue standard graph with linguistic annotation, but include one or several additional (typically) analogue modalities such as images or maps. Enhanced graphs go beyond the present discussion of unimodal modalities. They are mentioned here because of the importance of one of their forms, i.e. the 'data map' or 'thematic map' in which a reduced-specificity map assumes the role of graph space. Enhanced graphs are widely used to represent graph information to "graph illiterates", such as children. An enhanced bar graph is shown in Figure 10.

2.6. Beyond Literal Meaning. Metaphor and Metonymy

So far, we have focused only on the *literal meaning* of information representations. Thus, in interaction design, an image of an apple would be meant by the designer to refer to an apple or to apples, as the case may be; an acoustic image of people convening for a meeting would be meant to refer to people convening for a meeting; etc. However, using representational modalities in their literal meaning with the intention of their being understood as such during information representation and exchange is only one, albeit fundamental, form of information representation and hence of the use of representational modalities. Sometimes it may be preferable to use *non-literal meaning* instead, or in addition, i.e. to use modalities intending them to be understood in a way, which is different from their literal meaning. *Metaphoric use* of modalities is probably the best known kind of non-literal use in interaction design so far, such as in the static graphic desktop metaphor. What metaphors do is to bring a host of meaning and knowledge from a known source domain, such as the ordinary desktop, to bear on the user's understanding of the target domain, such as the computer screen. The trick is that the user knows the source domain already, so that simple and brief reference to that domain often suffices to marshal all of that understanding for the comprehension or reception of something new and unfamiliar. Metaphors are not the only kind of potentially useful non-literal meaning, however. We will consider only two kinds of non-literal meaning, i.e. metaphor and metonymy. In *metonymy,* a complex subject matter is being referred to through presenting some simple part of it.

In general, Modality Theory views non-literal meaning as being derived from literal meaning through subtraction of a smaller or greater amount of the literal connotations of an expression of information in some modality. Thus, the apple of MacIntosh Computers does not proclaim that you should buy, or use, an apple (Figure 7a). The claim is rather that you should get yourself something, which, like the apple shown, is related to many good things, such as knowledge, love, natural beauty and health. With respect to literalness, use of an apple to refer to computers is pretty far-fetched, coming close, but not quite there, to an arbitrary use of the representation of an apple. The metonymical chair used to refer to the program Director (Figure 7b) comes somewhat closer to literal meaning as the chair is one which is often associated with directing films.

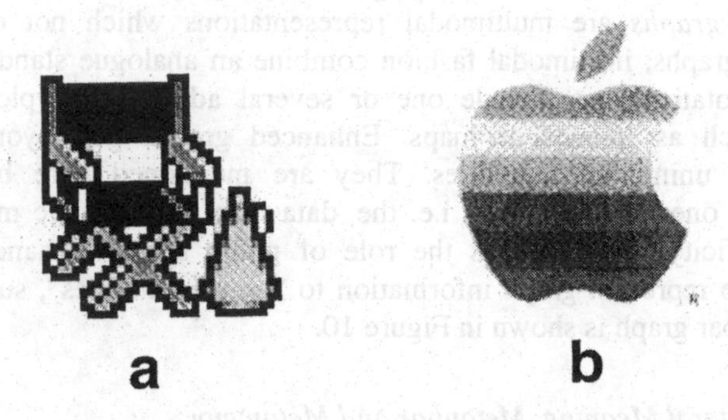

Figure 7. Metonymic (a) and metaphoric (b) representations.

In other words, beneath the surface of literal meaning, there is a deep in which any non-arbitrary expression of information in any modality can be used to potentially good effect in interaction design by making the expression convey non-literal information. At the bottom of this deep we find the arbitrary modalities, which have no (intended) relationship of meaning to that to which they refer. The idea is depicted in Figure 8. How to exploit the idea to good effect in interaction design is something, which, as usual, goes beyond Modality Theory.

In Section 2.3.2 we found that modalities can do more than just represent information. Modalities can be organised into *modality structures,* such as lists and tables, and modalities can assume *modality roles,* such as when being used as icons. In this section, we have seen that modalities can have *non-literal uses* in addition to, or in replacement of, their literal uses. These phenomena are not exclusive. It is perfectly possible, for instance, to make a table of metaphorical icons. Figure 8 is a case in point. The figure uses an explicit structure, abbreviated text, and image icons to metaphorically illustrate conceptual points made in the text.

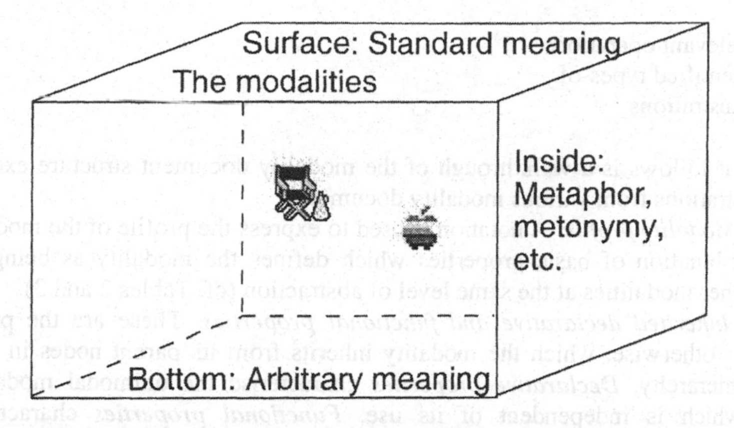

Figure 8. Non-literal uses of modalities can be located in the conceptual space between literal meaning and arbitrary meaning. The figure itself is a metaphor.

3. A FORMAT FOR REPRESENTING MODALITIES

Considered in isolation, the taxonomy of unimodal output modalities simply is a hierarchical analysis of the space of representational modalities in the media of graphics, acoustics and haptics. Gradually, as it were, the taxonomy turns into Modality *Theory* proper when (a) its generative principles are being accounted for in more detail, (b) its basic properties have been analysed in depth, (c) all individual unimodal modalities have been analysed as regards their properties and their capabilities and limitations for representing different types of information in context, and (d) other phenomena related to modality use, such as tables and metaphors, have found their proper place in the model. Points (a), (b), (c) and (d) have all been addressed above to some extent. Some time ago, we analysed all the unimodal modalities presented above and implemented them in a software demonstrator (Bernsen & Lu, 1995). We represented the analysis of each modality in a common format using a *modality document* template. Subsequent work on speech functionality (Section 5) has introduced some modifications, which have been included in the template below. The - still pending - completion of the taxonomy of unimodal input modalities may produce further revisions of the proposed common format for modality representation. This means that the following modality document template is a draft only.

Modality documents define, explain, analyse, and illustrate unimodal modalities from the point of view of interaction design support. The shared document structure includes the following entries:

(1) Modality profile
(2) Inherited declarative and functional properties
(3) Specific declarative and functional properties
(4) Combinatorial analysis

(5) Relevant operations
(6) Identified types-of
(7) Illustrations

What follows is a walkthrough of the modality document structure exemplified by illustrations from various modality documents.

(1) *Modality profile.* A notation is used to express the profile of the modality, i.e. the combination of basic properties which defines the modality as being distinct from other modalities at the same level of abstraction (cf. Tables 2 and 3).

(2) *Inherited declarative and functional properties.* These are the properties, basic or otherwise, which the modality inherits from its parent nodes in the taxonomy hierarchy. *Declarative properties* characterise the unimodal modality in a way, which is independent of its use. *Functional properties* characterise the unimodal modality as to which aspects of information it is good or bad at representing in context. An example is the claim that high-specificity (a large amount of detail in as many information channels as possible), high-image resolution, high-dimensionality (2 1/2D or 3D better that 2D) graphic images are useful for facilitating the visual identification of objects, processes, and events. An illustration of this functional property, and hence of one of the advantages of the static graphic image modality, is the use of photographs in criminal investigation. It is virtually impossible to linguistically express what a person looks like in such a way that the person may be uniquely identified from the linguistic description (Bernsen, 1995). Use of static graphic images, such as the one shown in Figure 11, can make this an effortless undertaking. Indeed, a picture can sometimes be worth more than a thousand words. Or, rather, this proverbial classic not only applies to pictures but to analogue representations in general, and irrespective of whether these are embodied in graphics, acoustics or haptics. The issue of functional properties of modalities is an extremely complex one, however, as we shall see (Section 5). For this reason, it is not entirely clear at present, which, if any, functional information should be included in individual modality documents. Except for the super level modalities, all unimodal modalities inherit important parts of their properties from higher levels in the taxonomy. A generic-level modality inherits the declarative and functional properties of its parent node at the super level; an atomic modality inherits the properties of its parent nodes at the super and generic levels, etc. To keep individual modality documents short, those properties should be retrievable through hypertext links. The following example shows the list of links to inherited properties in the atomic-level *gestural notation* modality document (Table 4) as well as the information channel and dimensionality information provided in the document.

- linguistic modalities
- static modalities
- dynamic modalities
- graphic modalities
- notation

- Static graphics have the following information channels: shape, size (length, width, height), texture, resolution, contrast, value (grey scales), colour (including brightness, hue and saturation), position, orientation, viewing perspective, spatial arrangement, short-duration repetitive change of properties.
- Dynamic graphics have the following information channels in addition to those of static graphics: non-short-duration repetitive change of properties, movement, displacement (relative to the observer), and temporal order.
- The dimensionality of dynamic graphics is 1D, 2D and 3D spatial, time.

The important analytical concept of an information channel may be briefly explained as follows. When, in designing human-human-system interaction, we choose a certain (unimodal) output modality to represent information, this modality inherits a specific medium of expression, such as acoustics, which it shares with a number of other unimodal modalities. An *information channel* is a perceptual aspect (an aspect accessible through human perception) of some *medium,* which can be used to carry information in context. If, for instance, differently numbered but otherwise identical iconic ships are being used to express positions of ships on a screen map, then different colourings of the ships can be used to express additional information about them. Colour, therefore, is an example of an information channel. A list of the information channels, which are characteristic of a particular unimodal modality, makes explicit the full inventory of means for information representation available to the developer using that modality.

Returning to the example above, gestural notation inherits the properties of the linguistic, static, dynamic, graphic, and notational modalities. As the information channel and dimensionality information is important to have close at hand, it is repeated in the document rather than having to be retrieved through hypertext links. Because of the pragmatic node-reduction (or fusion) strategy (Section 2.2), the gestural notation document presents both static and dynamic gestural notation. Figure 9 shows an example of static gestural notation.

(3) *Specific declarative and functional properties.* These are the properties which characterise the modality as being specifically different from its sister modalities with which it shares a common ancestry. For instance, in the *arbitrary modality* document (super level), the entry on 'Specific declarative and functional properties' includes the point that "Arbitrary modalities express information through having been defined ad hoc at their introduction." This implies that information represented in arbitrary modalities, whether graphic, acoustic or haptic, in order to be properly decoded by users, must be introduced in some non-arbitrary modality, such as some linguistic modality or other.

Figure 9. Static gestural notation: a marshalling signal which means 'move ahead'.

This is illustrated in Figure 10 in which ad hoc use of the graphic information channel colour (blue for the left-hand bar and green for the right-hand bar in a pair) has been defined in static graphic typed labels/keywords in the graph legend in the top right-hand corner. Without this linguistic annotation, it would not be possible to interpret the bar graph shown. The graph compares waste recycling of aluminium, glass and paper in the years 1970 and 1991 in the USA. An interesting point about the bar graph in Figure 10 is that the choice of the colours blue and green is not entirely arbitrary. In fact, the green colour used for 1991 manages to metaphorically suggest environmentalist progress. Metaphoric and other non-literal uses of modalities are discussed in Section 2.6.

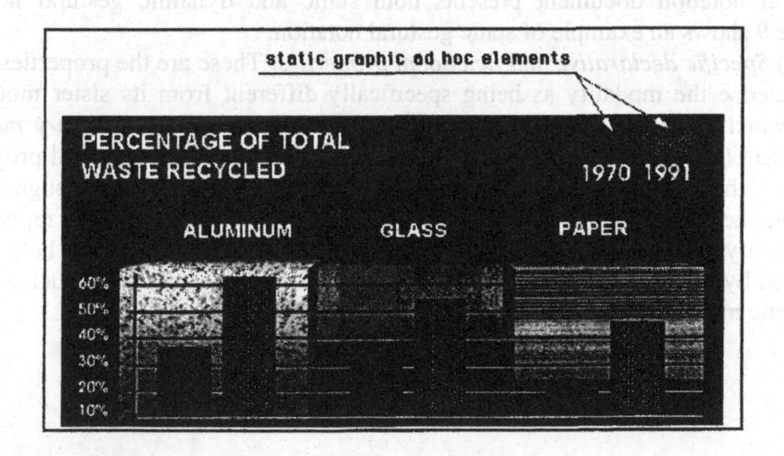

Figure 10. Dependence on linguistic modalities of an information channel used ad hoc.

(4) *Combinatorial analysis* expresses which other unimodal modalities a particular modality may or may not be combined with to compose multimodal representations. For instance, in the modality document on explicit static graphic structures the combinatorial analysis states that "explicit static graphic structures combine well with any static or dynamic graphic modality, whether linguistic, analogue or arbitrary". This is illustrated in Figure 12. The figure represents a Macintosh window as a layered series of unimodal explicit static graphic structures. Combinatorial analysis is highly important to the discovery of patterns of compatibility and incompatibility between unimodal modalities. Such patterns would begin to constitute a (unimodal) modality combination "grammar" or "chemistry". Like modality functionality analysis, combinatorial analysis faces a high-complexity problem space. We cannot claim as yet to have demonstrated a workable solution to the problem of how to do systematic combinatorial analysis (see Section 6).

Figure 11. A unimodal static graphic image of high specificity.

(5) *Relevant operations* are operations, which can be applied to the unimodal output modality described in a particular template. An operation may be defined as a meaningful addition, reduction or other change of information channels or dimensionality in a representation instantiating some modality. The purpose of an operation typically is to bring out more clearly particular aspects of the information to be presented. *Dimensionality reduction,* as in reducing common road maps from 3D to 2D without loss of key information; *specificity reduction,* as in replacing an image with a sketch; *saliency enhancement,* as in selective colouring; and *zooming* are some of the operations applicable to analogue graphic modalities. Similarly, **boldfacing**, *italicising*, underlining and re-sizing are common operations in graphic typed languages.

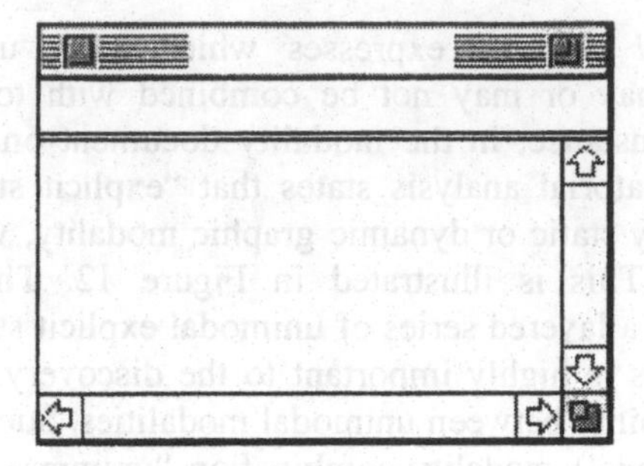

Figure 12. Nested unimodal explicit static graphic structures: the Macintosh window.

(6) *Identified types-of.* These are simply the sub-types of a unimodal modality, which are to be found one level down in the taxonomy hierarchy. In this way, the modality template presents an overview of the daughter modalities of the unimodal modality, which is currently being inspected. For instance, dynamic non-analogue sign graphic language (generic level) has six atomic types:

(1) Dynamic graphic written text
(2) Dynamic graphic written labels/keywords
(3) Dynamic graphic written notation
(4) Graphic spoken text or discourse
(5) Graphic spoken labels/keywords
(6) Graphic spoken notation

(7) *Illustrations.* Each modality document is illustrated by some 5-10 illustrations selected such as to show both prototypical examples, important non-prototypical or marginal cases, interesting multimodal combinations, non-literal uses, etc.

In addition to the systematic representation of all unimodal modalities in a common format, Modality Theory comes with a number of supporting theoretical concepts some of which have been mentioned already, such as 'specificity', 'interpretational scope', 'modality role', 'saliency', 'information channel', or 'dimensionality'. Theoretical concepts are explained and illustrated in *lexicon documents.* There are currently about 70 such documents (or concepts). Due to the heterogeneous nature of their topics, no rigid document structure has been imposed on lexicon documents.

4. TOWARDS A TAXONOMY OF UNIMODAL INPUT

The present section briefly describes ongoing work on a taxonomy of unimodal input modalities. We are almost, but not quite, there with a taxonomy of unimodal input modalities corresponding to the taxonomy of unimodal output modalities presented above. It may be illuminating to present and discuss some reasons why it has proved somewhat hard to arrive at a taxonomy of input modalities because, naively, one might have assumed that this would be merely a matter of mirroring the taxonomy of unimodal output modalities. Well, it is not.

It is interesting to note that the state of the art in input taxonomies is probably even less developed than the state of the art in output taxonomies. With respect to the latter, we have seen (Sections 1 and 2) that most existing taxonomies were confined to the static graphic modalities, reflecting the fact that static graphics antedate even the pyramids. The exception is the purely empirical output modality lists, which have been produced quite recently based on the multitude of studies of individual modality combinations during the 1990s.

Input, on the other hand, is in one sense a historical novelty because we did not have "real" input until the computer came along. In another sense, this picture is a rather superficial one, as we shall see. Not surprisingly, the first input taxonomies were proposed in the 1980s in an attempt to produce a scientific basis for addressing design choices among the growing number of different haptic input devices, such as the mouse, the joystick, the button, the pen, etc. The leading question, thus, was *not* one of distinguishing among modalities but of distinguishing among devices according to what a particular device was good or bad for. Some examples are Lenorovitz et al. (1984), who based their taxonomy of haptic input devices on *user goals,* such as 'create', 'indicate', 'eliminate', 'manipulate' and 'activate'. Foley et al. (1984) based their taxonomy of haptic input devices on basic input *interaction tasks,* such as 'select', 'position', 'orient', 'path', 'quantify' and 'text', and *control tasks,* such as 'stretch', 'sketch', 'manipulate' and 'shape'. Similarly, Greenstein & Arnaut (1988) distinguished between input tasks, such as 'point', 'draw', 'trace' and 'track'. Other researchers in this line of research are, e.g., Buxton (1983) and Mackinlay et al. (1990). Their common goal was to create a systematic basis of rules or guidelines of the form: "if the task (or user goal) is Tx (or UGy), then use haptic device HDn". Good systematic work was done towards these aims. Nevertheless, this line of research seems to have disappeared in the early 1990s. The reasons why this happened are interesting and important. The researchers gradually realized that they were building on extremely unstable foundations. The fact is that we still do not have any stable taxonomies of user goals (as it happens, psychologists already gave up on that in the 1940s) or user tasks. Moreover, we do not seem likely to have such taxonomies in the foreseeable future, as will be argued in Section 5. But if those taxonomies are unattainable for practical purposes, prospects are dim to ever reach any kind of systematicity or closure on sets of rules or guidelines of the form described above. Another fact of the matter is that new devices get invented at increasing speed, leaving any device taxonomy gasping for breath to catch up and revise its user goals or task type foundations post hoc. This is a problem for any attempt at creating a science-based device taxonomy.

Other than the haptic input taxonomies just described, few attempts appear to have been made, and none appear to have been made to gain a systematic grasp of input *modalities.* The best we have are lists of input devices, modalities and tasks based on the literature from the 1990s (e.g. Benoit et al., 2000), and focused sets of contributions towards understanding the pros and cons of particular modalities and multimodal combinations but lacking in attempts at systematic theoretical comprehension. A good example is the volume of articles on speech functionality edited by Baber & Noyes (1993).

Input modalities are forms of representation of information from the user to the machine. When we began work on a taxonomy for unimodal input modalities, a natural first question was whether it might be possible to simply re-use the output taxonomy or whether the output taxonomy would have to be modified to account for input. Closer analysis revealed a number of – real or purported - asymmetries between output and input modalities. The ones identified so far are discussed in the following.

4.1. Asymmetries between Output and Input

4.1.1. Perceptual asymmetry

Obviously, interactive *output* must be perceptible to humans in HHSI. This is not necessarily true for input where it is sufficient that the machine is able to perceive what arrives at its sensors. Thus, some input media aspects are not (fully) perceptible to humans, such as radar, infrared, ultrasound, magnetic fields, skin conductivity, etc. An implication is that our list of output media *information channels* is likely to have to be augmented when dealing with input.

4.1.2. Media asymmetry

It is an obvious fact that haptics is, at least so far, much more prominent in input than in output. As far as output is concerned, current systems can output information in any output modality as long as we have plenty of time to create the output at design-time, and as long as we have built the output devices we need. In input, haptics dominate. We key in text, point, track and click with the mouse, write and draw with the pen, manipulate 3D objects to modify graphical output representations, move in force-feedback output space, etc. Correspondingly, as we saw above, most input theory is about haptic input, focusing on the haptic control and/or creation process. There are at least two reasons why this is the case. The first reason is that humans are very good at using their hands, which is reflected in the fact that most of the control devices, which have been invented before and after the industrial revolution are haptic devices. The second reason is that acoustic (including speech) recognition and understanding by machine has been difficult to achieve and that visual recognition and understanding by machine is more difficult still. Workable speech recognition systems have been around for about two decades only, and workable spoken dialogue systems have been around for about a decade only (Bernsen et al., 1998). It is not surprising, therefore, that haptic devices continue to proliferate.

Based on the above observations, the following claim would appear plausible: we should forget about (most of) *device theory* and focus on Modality Theory instead. As was pointed out earlier, haptic device theory became extinct because scientific soundness was missing. There are no formally sound taxonomies of user goals or tasks in HHSI. Modality theory has stronger formal properties, at least as concerns output. Secondly, device theory would seem destined to become much too simplistic for the non-haptic input media. All we need as regards input devices for acoustics and graphics are microphones, cameras, infrared sensors, and a few more. Even factoring in more detailed distinctions among technologies, such as between direction-sensitive microphones, microphone arrays and the like, device theory is not likely to attain the level of detail which is needed to support interaction design. Thirdly, input devices are generally easy to build and modify, which is also why it is virtually impossible to keep track on developments. In fact, there is something deeply puzzling about the haptic input device theory endeavor, something which we have termed the *doo knob problem.* Who ever saw the ultimate textbook on doorknobs, or on knives? Mankind has been using doorknobs for centuries and knives for much longer, yet the literature on doorknobs or knives, the taxonomies of them, the proposals for scientific foundations for judging the appropriateness or inappropriateness of various types of knives and doorknobs for particular purposes, etc., is non-existent. The simple reason appears to be that if a particular "type" of knife or doorknob does not work for a particular purpose, we just design and manufacture a better one. The same holds, or will soon hold, it is suggested, for haptic input devices. So, who needs device theory? Or, re-phrased from the point of view of Modality Theory, let us start with the information to be exchanged in HHSI and then find, or make, the input devices, which can meet the requirements.

4.1.3. Is input "more essentially interactive" than output?

When investigating possible asymmetries between output and input, we also came across more questionable purported asymmetries than the above, such as the following.

If we focus on what people *do* with systems during HHSI, it is tempting to conclude that we mostly just *receive* ready-made output but that we often *create* things with the input we produce, such as text, drawings, 3D graphics objects, soundscapes, a flight ticket reservation made through spoken input, etc. We are less accustomed to the point of view that machines also do something to, or with, humans during interaction. Thus, the output taxonomy presented above has focused on ready-made results, such as the images or the music to be presented as system output, rather than on the process of interactively creating those output modalities. However, when considering input, it is much harder to ignore the *input process* through which we create those (output) representations. In other words, *input is input into an output space,* which gets modified as a result of the input. With today's WYSIWYG and direct manipulation interfaces, we have become accustomed to getting ample immediate feedback on the results of our input actions.

It is less easy to accept that *output is output into an input space* (i.e. the user's cognitive system), which gets modified as a result of the output. Nevertheless, this is how the *developer* always had to conceive of output creation, i.e. that the output is there to modify the cognitive state and behavior of the humans interacting with the system. To be meaningful in context, both input and output require knowledge of the interlocutor's state and how it is likely to change as a result of the information, which is being sent across. An example in point is spoken language dialogue systems output design during which the developer must constantly refer to how the user is likely to understand the system's output (Bernsen et al., 1998). With increasingly advanced (or more intelligent) machines able to do flexible and adaptive interaction on-line, this burden will gradually shift from the developer to the system itself. In other words, so far, users need feedback more than machines do; so far, users tend to work into an output environment more than machines tend to work into an input environment on-line; so far, machines mostly produce "finished things" on-line; and so far, machines are tools for humans more than humans are "tools" for machines. But this only goes to say that, so far, machines are (mostly) dumber than humans in their perception and understanding of their interlocutors during HHSI. And this will gradually change.

4.1.4. Dynamic/static modalities asymmetry

An important difference between output and input seems to be that the dynamic/static distinction does not apply to machines. It does not matter to the machine whether the input it receives is static or dynamic in the sense of these terms used in Modality Theory (Section 2.1). The reason is that machines are endowed with "photographic memory", as a result of which they tend to capture all the information contained in the input, within, of course, the capabilities of their input sensors. Once captured, the machine does not profit from, or need, having exactly the same input repeated. Instead, they use their processing power to internally exploit the information already received. Humans are different. They do profit from repetition of complex information, such as a static graphics screenful of information, just as they profit from the constant availability of complex information whilst they make up their minds on what to do next. This is why the static modalities are so important to humans in many cases, and why humans have difficulties receiving complex information dynamically. Nothing prevents us from building machine, which share with humans the property of selective focusing within a complex information space, machines which quickly forget most of the complex information they are being exposed to and which have to refresh their memory from time to time by re-perceiving the information with a changing perceptual and cognitive focus. So far, however, there seems to be little point in building such machines. Removal from the unimodal input taxonomy of the static/dynamic distinction significantly simplifies input taxonomy generation. For input, the basic matrix in Table 1 reduces to 24 initial modalities (Table 10).

4.1.5. Conclusion

In conclusion, when generating the taxonomy of unimodal input modalities, we need to watch out for input which is perceptible to machines but not perceptible to humans. We will probably have to expand the haptic section of the input taxonomy compared to that in the output taxonomy given the importance of haptics in human-to-machine interaction. The problem, however, in expanding on input haptics is that this should not be done through reference to scientifically unsound taxonomies of user goals, tasks or otherwise. Similarly, we should continue to resist the temptation to involve (input) device properties in the taxonomy even if those properties are sometimes important to device selection. Rather, if none of the devices available are appropriate, we build a new and better (faster, more precise, etc.) one. Finally, we should omit the static/dynamic distinction in the input taxonomy.

Table 10. The full set of 24 combinations of basic properties constituting the possible unimodal input modalities at the generic level of the taxonomy. All modalities provide possible ways of representing information.

	li	*-li*	*an*	*-an*	*ar*	*-ar*	*gra*	*aco*	*hap*
1	x		x		x		x		
2	x		x		x			x	
3	x		x		x				x
4	x		x			x	x		
5	x		x			x		x	
6	x		x			x			x
7	x			x	x		x		
8	x			x	x			x	
9	x			x	x				x
10	x			x		x	x		
11	x			x		x		x	
12	x			x		x			x
13		x	x		x		x		
14		x	x		x			x	
15		x	x		x				x
16		x	x			x	x		
17		x	x			x		x	
18		x	x			x			x
19		x		x	x		x		
20		x		x	x			x	
21		x		x	x				x
22		x		x		x	x		
23		x		x		x		x	
24		x		x		x			x
	li	*-li*	*an*	*-an*	*ar*	*-ar*	*gra*	*aco*	*hap*

4.2. Towards a taxonomy of unimodal input modalities

Given the argument above, the first step of generating a complete matrix of unimodal input modalities at the generic level is easy (Table 10). Removing the modalities which represent the arbitrary use of information tokens which already have an established meaning is simple as well (Table 11). We are currently experimenting with the generation of the atomic level of unimodal input modalities. The way this is being done is to use questionnaires in the form of tables showing an atomic-level breakdown of the generic level into intuitive and relevant atomic unimodal input modalities. Subjects are asked to fill in each cell of the table with one or more concrete examples in which the atomic input modality in point is being produced interactively into a particular output space and using specific input devices. This process generates issues of classification and consistency with the atomic output taxonomy, which are then being analyzed and discussed. For illustration, some examples from the questionnaires follow.

Table 11. 15 generic unimodal input modalities result from removing from Table 10 the arbitrary use of non-arbitrary modalities of representation.

	li	*-li*	*an*	*-an*	*ar*	*-ar*	*gra*	*aco*	*hap*
1	x		x			x	x		
2	x		x			x		x	
3	x		x			x			x
4	x			x		x	x		
5	x			x		x		x	
6	x			x		x			x
7		x	x			x	x		
8		x	x			x		x	
9		x	x			x			x
10		x		x	x		x		
11		x		x	x			x	
12		x		x	x				x
13		x		x		x	x		
14		x		x		x		x	
15		x		x		x			x
	li	*-li*	*an*	*-an*	*ar*	*-ar*	*gra*	*aco*	*hap*

(1) Linguistic graphics: Written notation: A mathematician in front of a blackboard filled with formulae tells the system: "Please digitise and store!"

(2) Analogue graphics: Maps: A person shows the system a map and asks for the nicest itinerary to a particular landmark.

(3) Analogue acoustics: Graphs: Singing people compete to produce overtones in Italian arias. The system is the judge.

(4) Analogue haptics: Manipulation/action: A person explores a VR cityscape on a bicycle.

5. APPLYING MODALITY THEORY TO SPEECH FUNCTIONALITY

Following the research agenda of Modality Theory, we should at this point address the issue of how to combine unimodal output modalities, unimodal input modalities, and unimodal input/output modalities into usable multimodal representations. However, as the taxonomy of unimodal input modalities is not quite ready yet, this issue will be postponed to the final section of the present chapter. This section briefly addresses the two final issues on the research agenda of Modality Theory, i.e. (4) to develop a methodology for applying Modality Theory to early design analysis of how to map from the requirements specification of some application to a usable selection of input/output modalities, and (5) to use results in building, possibly automated, practical interaction design support tools. These two issues have been the subjects of substantial work for several years already, resulting in a number of system design case studies, comprehensive studies of speech functionality, and a design support tool. Before describing the progress made, it may be useful to review some false starts on methodology, which, as was discovered in the process, we were not the only ones to make.

5.1. How (not) to Map from Requirements Specification to Modality Selection

As a result of a series of multimodal systems design case studies, a consolidated methodology for mapping from requirements specification to modality selection was specified (Bernsen & Verjans, 1997). The basic notion of the methodology was that of *information mapping rules*. The modality document template (cf. Section 3) would list, per unimodal output or input modality, the information mapping rules applying to that modality. The rules would then be used, manually at first but eventually automatically by a rule-based system, to generate advice on which modalities to use in designing interaction for a particular application. The final product would be an 'interface sketch' which identifies the best input and output modalities to use, the devices to use, and the interactive functionalities which the artefact to be developed would need. All the sketch would need to become a full design specification was a more detailed task analysis together with the application of a particular standard or aesthetic design concept for controlling the detailed specification of the interface. As it turned out, the inherent complexity of the problem space of selecting among thousands of potential modality combinations subject to multiple constraints imposed by the context of use of the artefact to be developed, proved too great for a rule-based approach. We were envisioning getting bogged down by thousands of rules expressed in a basically unsound and non-transparent state of the art conceptual apparatus for describing tasks, user groups, user goals, and many other parameters as well. We also discovered that other researchers, such as the group in Namur, Belgium (e.g. Bodart et al., 1995), had spent years working on a small (static data graphics) fraction of the modalities covered by Modality Theory, only to arrive at the same adverse conclusion. In other words, we had to abandon methodology A below.

A. Rule-based mapping onto modality selection:

(1) Initial requirements specification ->
(2) rules specifying what a particular modality can be used for ->
(3) interface sketch and device selection ->
(4) detailed task analysis and interface specification.

In retrospect, this methodology shares the flaws of the haptic input taxonomy work reviewed in Section 4, i.e. methodology B below.

B. Rule-based mapping onto device selection:

(1) Requirements specification ->
(2) task/goal taxonomy + rules ->
(3) mapping onto device selection.

After these lessons in infeasibility, it was clear that any solution to the mapping problem would have to be based on a significant reduction in complexity. We also knew that some of the possible complexity reductions would not be sufficient. Thus, merely reducing complexity to a sub-section of unimodal modality space would not help. We would still be bogged down by scientifically unsound, as well as unmanageable, rules. So, the rule-based approach common to Methodologies A and B above had to go. Still, reducing complexity to a sub-section of unimodal modality space might at least help us experiment with new methodologies without having to take on all unimodal modalities and their thousands of multimodal combinations from the outset. We decided to consider only speech output and speech input together with multimodal representations, which include speech. During the study of the literature on speech functionality, i.e. on the issue of when it is (not) advisable to use speech output and/or speech input in interaction design, it became clear that there is a crucial distinction between the rules involved in the rule-based approaches A and B above, and principles which simply state declarative properties (cf. Section 3) of unimodal modalities. A rule would state something like:

If the task is Tl and the user group is UGm and the goal is to optimise efficiency of interaction, then use modality (or modality combination) Mn.

The problem, as we have seen, is the scientifically messy notions of task types, user group types and performance parameter types (such as efficiency). Moreover, there is not even consensus about the number and nature of the *types* of relevant parameters. On the other hand, the modalities themselves and their declarative characteristics *is*, potentially at least, a scientifically sound part of a possible methodology. An example of a declarative characteristic of a modality is:

Speech is omnidirectional.

We call such characteristics *modality properties*. The methodology (C below), then, is based on modality properties rather than rules.

C. Modality property-based mapping onto modality selection:

(1) Requirement specification ->
(2) modality properties + natural intelligence ->
(3) advice/insight with respect to modality choice.

Clearly, methodology C is considerably more modest than methodologies A and B above. In particular, C is non-automated and relies on human intelligence for deriving clues for modality choice decisions from modality properties. All we can do, it would seem, to help train and sharpen the human intelligence doing the derivations, is to provide a wide range of concrete interaction design examples, each of which has been analysed with respect to the modality choice claims or decisions made. This we set out to do for the speech functionality problem.

5.2. Speech Functionality

In two studies, we have analysed in depth 273 claims about speech functionality found in the literature on the subject from 1993 onwards (Bernsen, 1997; Bernsen & Dybkjær, 1999a). The claims were selected in an objective fashion in order to prevent exclusion of examples, which might prove difficult to handle through Methodology C (see Section 5.1). Thus, the first set of 120 claims studied was the entire set of claims to be found in Baber & Noyes (1993). The second set of claims was identified from the literature on speech functionality since 1993 by a researcher other than the one who would be analysing the set. The exercise has provided a solid empirical foundation for judging just how complex the speech functionality problem is in terms of the number of instantiated parameters involved (the modality complexity we knew already, more or less). The complexity of the problem of accounting for the functionality of speech in interaction design is apparent from the semi-formal expression from (Bernsen & Dybkjær, 1999b) in Figure 13. Parameters are in boldface.

[Combined speech input/output, speech output, or speech input modalities M1, M2 and/or M3 etc.] or
[speech modality M1, M2 and/or M3 etc. in combination with non-speech modalities NSM1, NSM2 and/or NSM3 etc.] are [useful or not useful] for:
[**generic task** GT and/or **speech act type** SA and/or **user group** UG and/or **interaction mode** IM and/or **work environment** WE and/or **generic system** GS and/or **performance parameter** PP and/or **learning parameter** LP and/or **cognitive property** CP] and/or are:
[preferable or non-preferable] to:
[alternative modalities AM1, AM2 and/or AM3 etc.] and/or are:
[useful on conditions] C1, C2 and/or C3 etc.

Figure 13. Minimal complexity of the speech functionality problem.

Note that each of the boldfaced parameters can be instantiated in many different ways. For instance, Bernsen (1997) found 38 different instantiations of the parameter *performance parameter* in the 120 claims analysed. It is a sobering thought that any systematic approach to modality choice support must face this complex parameter space in addition to the complexity of the space of unimodal input/output modalities and their combinations.

Prior to the claims analysis, each of the 273 claims from the literature was rendered in a standard semi formal notation in order to facilitate analysis. Each rendering-cum-analysis had to be independently approved by a second researcher. An example of a rendering-cum-analysis is Claim (or data point) 48 in Figure 14, quoted from (Bernsen & Dybkjær, 1999b). The claim representation first quotes the original expression of the claim followed by a literature reference. The claim is then expressed in a standard format referring to the modalities and parameters involved.

The statement "Justified by MP5" refers to one of the modality properties used in evaluating the claims (Table 12). The claims evaluation is the centerpiece of the work reported here. For each claim, search was made among the hundreds of modality properties of Modality Theory to identify those properties, which could either justify, support (but not fully justify), or correct the claim. In the case of Claim 48, as shown, one modality property was found which justifies the claim. The Claims type "Rsc" refers to claims, which recommend the use of speech in combination with other modalities. The claims evaluation in Figure 14 is followed by an (optional) note on the claim and its evaluation. Finally, the claim itself is evaluated as being true. This evaluation of the truth-value of claims is important, because the potential of modality properties for claims analysis is basically to be judged by the percentage of true claims which Modality Theory, through reference to modality properties, is able to justify or support, and the percentage of false or questionable claims which the theory is able to correct. It would be bad news for the approach if there were many true, questionable, or false claims on which the theory had nothing to say.

48. Interfaces involving spoken ... input could be particularly effective for interacting with dynamic map systems, largely because these technologies support the mobility [walking, driving etc.] that is required by users during navigational tasks. [14, 95]
Data point 48. **Generic task** [mobile interaction with dynamic maps, e.g. whilst walking or driving]: a speech input interface component could be **performance parameter** [particularly effective].
Justified by MP5: "Acoustic input/output modalities do not require limb (including haptic) or visual activity." Claims type: **Rsc.**
NOTE: The careful wording of the claim "Interfaces involving spoken ... input". It is not being claimed that speech could suffice for the task, only that speech might be a useful interface ingredient. Otherwise, the claim would be susceptible to criticism from, e.g., MP1. Note also that the so-called "dynamic maps" are static graphic maps, which are interactively dynamic.
True.

Figure 14. Evaluation of a claim about speech functionality.

What we found, however, was that Modality Theory was able to justify, support, or correct 97% of the claims in the first study of 120 claims (Bernsen, 1997), and 94% of the claims in the second study of 153 claims (Bernsen & Dybkjær 1999a). In other words, assuming, as argued in those two studies, the representativity of the analysed claims with respect to all possible claims about speech functionality, modality properties – i.e. the clean, declarative messages of Modality Theory – are highly relevant to judging speech functionality in early design and development.

A final important question is: how many modality properties were needed to achieve the high percentages reported in the preceding paragraph? In fact, the first

Table 12. Modality properties found relevant to speech functionality evaluation.

No	Modality	MODALITY PROPERTY
MP1	Linguistic input/output	Linguistic input/output modalities have interpretational scope, which makes them eminently suited for conveying abstract information. They are therefore unsuited for conveying high-specificity information including detailed information on spatial manipulation and location.
MP2	Linguistic input/output	Linguistic input/output modalities, being unsuited for specifying detailed information on spatial manipulation, lack an adequate vocabulary for describing the manipulations.
MP3	Arbitrary input/output	Arbitrary input/output modalities impose a learning overhead which increases with the number of arbitrary items to be learned.
MP4	Acoustic input/output	Acoustic input/output modalities are omnidirectional.
MP5	Acoustic input/output	Acoustic input/output modalities do not require limb (including haptic) or visual activity.
MP6	Acoustic output	Acoustic output modalities can be used to achieve saliency in low-acoustic environments. They degrade in proportion to competing noise levels.
MP7	Static graphics/haptics input/output	Static graphic/haptic input/output modalities allow the simultaneous representation of large amounts of information for free visual/tactile inspection and subsequent interaction.
MP8	Dynamic input/output	Dynamic input/output modalities, being temporal (serial and transient), do not offer the cognitive advantages (wrt. attention and memory) of freedom of perceptual inspection.
MP9	Dynamic acoustic output	Dynamic acoustic output modalities can be made interactively static (but only small-piece-by-small-piece).
MP10	Speech input/output	Speech input/output modalities, being temporal (serial and transient) and non-spatial, should be presented sequentially rather than in parallel.
MP11	Speech input/output	Speech input/output modalities in native or known languages have very high saliency.
MP12	Speech output	Speech output modalities may complement graphic displays for ease of visual inspection.
MP13	Synthetic speech output	Synthetic speech output modalities, being less intelligible than natural speech output, increase cognitive processing load.
MP14	Non-spontaneous speech input	Non-spontaneous speech input modalities (isolated words, connected words) are unnatural and add cognitive processing load.
MP15	Discourse input/output	Discourse input/output modalities have strong rhetorical potential.
MP16	Discourse input/output	Discourse input/output modalities are situation-dependent.
MP17	Spontaneous spoken labels/keywords and discourse input/output	Spontaneous spoken labels/keywords and discourse input/output modalities are natural for humans in the sense that they are learnt from early on (by most people and in a particular tongue and, possibly, accent). (Note that spontaneous keywords and discourse must be distinguished from designer-designed keywords and discourse which are not necessarily natural to the actual users.)
MP18	Notational input/output	Notational input/output modalities impose a learning overhead which increases with the number of items to be learned.

MP 19	Analogue graphics input/output	Analogue graphics input/output modalities lack interpretational scope, which makes them eminently suited for conveying high-specificity information. They are therefore unsuited for conveying abstract information.
MP 20	Haptic manipulation selection input	Direct manipulation selection input into graphic output space can be lengthy if the user is dealing with deep hierarchies, extended series of links, or the setting of a large number of parameters.
MP 21	Haptic deixis (pointing) input	Haptic deictic input gesture is eminently suited for spatial manipulation and indication of spatial location. It is not suited for conveying abstract information.
MP 22	Linguistic text and discourse input/output	Linguistic text and discourse input/output modalities have very high expressiveness.
MP 23	Images input/output	Images have specificity and are eminently suited for representing high-specificity information on spatio-temporal objects and situations. They are therefore unsuited for conveying abstract information.
MP 24	Text input/output	Text input/output modalities are basically situation-independent.
MP 25	Speech input/output	Speech input/output modalities, being physically realised in the acoustic medium, possess a broad range of acoustic information channels for the natural expression of information.

study needed only 18 modality properties whilst the second study needed seven additional modality properties. It may thus be concluded that a relatively small number of modality properties constitute an extremely powerful resource for evaluating most speech functionality claims or assumptions in early design and development. The modality properties used in the two studies are listed in Table 12.

5.3. The SMALTO Tool

The results presented in Section 5.2 convinced us that it might be worthwhile to develop a design support tool for supporting early design decisions on whether or not to use speech-only or speech in multimodal combinations for particular applications. The tool is called SMALTO and can be accessed at http://disc.nis.sdu.dk/smalto/. Basically, what SMALTO does is to enable hypertext navigation among hundreds of evaluated claims made in the literature as well as among the modality properties, which bear on those claims. The benefits to be derived from using the tool are to become familiar with the specific modality thinking which bears on the design task at hand in case claims are found which are immediately relevant to that design task, and to become increasingly familiar with the general modality thinking which can be done straight from an understanding of the modality properties themselves (Luz & Bernsen, 1999).

6. MULTIMODALITY

Getting a theoretical handle on multimodality would constitute a major result of Modality Theory. As this is work in progress, we are not yet able to present any well-tested approach, which could be claimed to be superior to, or a valuable complement to, the best current approach.

6.1. The Best Current Approach

The best current approach to the issue of multimodality as described in this chapter is an empirical one. The approach consists in, quite simply, analysing and publicising "good compounds", i.e. good modality combinations with added observations on the parameter instantiations which have been studied or which are being hypothesised about (cf. Section 5.2). As has become clear from the argument in this chapter, no modality combination is good for every conceivable purpose and it is difficult, to say the least, to precisely circumscribe the circumstances in which a particular modality combination should be preferred to others for the representation and exchange of information. Still, it is useful to build systematic overviews of modality combinations, which have proved useful for a broad range of specified purposes. Some such combinations have already been described above, such as bimodal static graphics including, e.g., linguistically labelled graphs and static graphics images, which illustrate static graphic text. There are very many other "good compounds", such as speech combined with static or dynamic graphics output, speech and pen input, speech and analogue haptics output for the blind, etc. At least, when using a modality combination which has been certified as a good one under particular circumstances, developers will know that they are not venturing into completely unexplored territory but can make the best use of what is already known about their chosen modality combination. The problems about this approach are, first, that it tends to be a rather conservative one, dwelling upon modality combinations which have been used so often already that some kind of incomplete generalisation about their usefulness has become possible. Thus, the approach lacks the predictive, creative and systematic powers of a firm theoretical grasp of the space of possible modality combinations. Secondly, given the complex parameter space addressed by any claim about the usefulness of a particular modality combination, most of those generalisations are likely to be scientifically unsound ones, and increasingly so the more general they are.

6.2. Modality Theory–Based Approaches

How might Modality Theory do better than the best current approach (Section 6.1)? The theory is superior to that empirical approach in that Modality Theory allows a complete *generation* of all possible input, output, and input/output modality combinations at any level, such as the atomic level, and cross level as well. However, whilst complete generation is possible in a way that is sufficient for all practical interaction design purposes, the combinatorial explosion involved makes it practically impossible to systematically *investigate* all the generated modality combinations. For instance, if we wanted to investigate all possible n-modal atomic input/output modality combinations where n=10, the number of combinations to be investigated would run into millions. Still, there do seem to be interesting opportunities for exploring this generative/analytic approach by carving out *combinatorial segments* from the taxonomy for systematic analysis, such as a speech-cum-other-modalities segment, or an input-manipulation-cum-other-modalities segment. These exercises could be further facilitated by tentatively clustering families of similar modalities and treating these as a single modality whose interrelations with other modalities are being investigated. An example could be to treat all analogue static graphics modalities as a single modality given the fact that these modalities have a series of important properties in common. It is perfectly legitimate to ask questions, such as "How does this particular family of tri-modal combinations combine with other modalities?"

An alternative to the generative approach just described could be to *scale up the SMALTO tool* to address all possible modality combinations. The problem, however, is that this would be likely to produce lists of hundreds of relevant modality properties, creating a space of information too complex for practical use. Part of the usefulness of SMALTO lies in the fact that SMALTO operates with such a small number of modality properties that it is humanly possible to quickly achieve a certain familiarity with all of them, including the broad implications for interaction design of each of them. It might be preferable, therefore, to use the SMALTO approach in a slightly different way, i.e. by producing modality properties for limited segments of combinatorial input/output modality space according to current needs just like SMALTO itself does.

We are currently working on a third approach which is to *turn Modality Theory as a whole into a hypertext/hypermedia tool* using a common format for modality representation similar to the format described in Section 3 but with an added entry for modality properties. By definition, the tool would include all identified modality properties. The challenge is to make the tool useful for interaction designers who are not, and do not want to become, experts in the theory, for instance by including a comprehensive examples database. In itself, this tool would not constitute a full scientific handle on multimodality in the sense of a systematic approach to multimodal combinations. However, building the Modality Theory tool does seem to constitute a necessary next step, which would also facilitate achievement of the ultimate goal of mastering the huge space of multimodal combinations.

Finally, a fourth approach is to *analyse the "good compounds"* (Section 6.1) in terms of modality properties in order to explore whether any interesting scientific generalisations might appear.

6.3. Conclusion

This chapter has outlined the current state of progress on the research agenda of Modality Theory. So far, Modality Theory has taken a different route from most recent work on modalities, which has focused on exploring, on an ad hoc basis, useful modality and/or device combinations based on an emerging conceptual apparatus, including concepts such as modality 'complementarity' and 'redundancy'. Modality Theory, on the other hand, has focused on generating fundamental concepts and taxonomies of unimodal output and input modalities subject to the requirements of completeness, uniqueness, relevance, and intuitiveness; on exploring methodologies for applying the emerging theory in design and development practice; and on developing demonstrator tools in support of modality choice decision making in early design of human-human-system interaction. In the process, a good grasp has been achieved of the extreme complexity of the problem of modality functionality. To complete the research agenda of Modality Theory, we need a well-tested taxonomy of unimodal input modalities, a Modality Theory hypertext/hypermedia tool, and exploration of additional ways in which the theory can be of help in achieving a systematic, creative, and predictive understanding of input/output modality combinations, including those which have not yet been widely used, if at all. These themes are topics of current research, which the reader is kindly invited to join.

7. REFERENCES

Baber, C. & J. Noyes (Eds.). *Interactive Speech Technology*. London: Taylor & Francis, 1993.

Benoit, C., J.C. Martin, C. Pelachaud, L. Schomaker & B. Suhm. "Audio-Visual and Multimodal Speech Systems." In: D. Gibbon (Ed.), *Handbook of Standards and Resources for Spoken Language Systems - Supplement Volume*. Kluwer, 2000.

Bernsen, N.O. "A research agenda for modality theory." In: Cox, R., Petre, M., Brna, P., and Lee, J. (Eds.), *Proceedings of the Workshop on Graphical Representations, Reasoning and Communication.* World Conference on Artificial Intelligence in Education. Edinburgh, 1993: 43-46.

Bernsen, N.O. "Foundations of multimodal representations. A taxonomy of representational modalities." *Interacting with Computers* 6.4 347-71, 1994.

Bernsen, N.O. "Why are analogue graphics and natural language both needed in HCI?" In: Paterno, F. (Ed.), *Design, Specification and Verification of Interactive Systems. Proceedings of the Eurographics Workshop,* Carrara, Italy, 165-179. *Focus on Computer Graphics.* Springer Verlag, 1995: 235-51, 1994.

Bernsen, N.O. "Towards a tool for predicting speech functionality." *Speech Communication* 23: 181-210, 1997.

Bernsen, N.O. "Natural human-human-system interaction." In: Earnshaw, R., R Guedj, A. van Dam & J. Vince (Eds.). *Frontiers of Human-Centred Computing, On-Line Communities and Virtual Environments.* Berlin: Springer Verlag, 2000.

Bernsen, N.O. & L. Dybkjær. "Working Paper on Speech Functionality." *Esprit Long-Term Research Project DISC Year 2 Deliverable* D2.10. University of Southern Denmark. See www.disc2.dk, 1999a.

Bernsen, N.O. & L. Dybkjær. "A theory of speech in multimodal systems." In: Dalsgaard, P., C.-H. Lee, P. Heisterkamp & R. Cole (Eds.). *Proceedings of the ESCA Workshop on Interactive Dialogue in Multi-Modal Systems,* Irsee, Germany. Bonn: European Speech Communication Association: 105-108, 1999b.

Bernsen, N.O., H. Dybkjær & L. Dybkjær. *Designing Interactive Speech Systems. From First Ideas to User Testing.* Springer Verlag, 1998.

Bernsen, N.O. & S. Lu. "A software demonstrator of modality theory." In: Bastide, R. & P. Palanque (Eds.). *Proceedings of DSV-IS'95: Second Eurographics Workshop on Design, Specification and Verification of Interactive Systems.* Springer Verlag, 242-61, 1995.

Bernsen, N.O. & S. Verjans. "From task domain to human-computer interface. Exploring an information mapping methodology." In: John Lee (Ed). *Intelligence and Multimodality in Multimedia Interfaces.* Menlo Park, CA: AAAI PressURL: http://www.aaai.org/Press/Books/Lee/lee.html, 1997.

Bertin, J. *Semiology of Graphics. Diagrams. Networks. Maps.* Trans. by J. Berg. Madison : The University of Wisconsin Press, 1983.

Bodart, F., A.M., Hennebert, J.-M. Leheureux, I. Provot, G. Zucchinetti & J. Vanderdonckt. "Key Activities for a Development Methodology of Interactive Applications." In: Benyon, D. & P. Palanque (Eds.). *Critical Issues in User Interface Systems Engineering,* Springer Verlag, 1995.

Buxton, W. "Lexical and pragmatic considerations of input structures." *Computer Graphics* 17,1: 31-37, 1983.

Foley, J.D., V.L., Wallace & P. Chan. "The Human Factors of Graphic Interaction Techniques." *IEEE Computer Graphics and Application* 4.11: 13-48, 1984.

Greenstein, J.S. & L.Y. Arnaut. "Input devices." In: M. Helander (Ed.). *Handbook of Human-Computer Interaction,* Amsterdam: North-Holland, 495-519, 1988.

Holmes, N. *Designer's Guide to Creating Charts and Diagrams.* New York: Watson-Guptill Publications, 1984.

Hovy, E. & Y. Arens. "When is a picture worth a thousand words? Allocation of modalities in multimedia communication." Paper presented at the *AAAI Symposium on Human-Computer Interfaces,* Stanford, 1990.

Joslyn, C., C. Lewis & B. Domik. "Designing glyphs to exploit patterns in multidimensional data sets." *CHI'95 Conference Companion,* 198-199, 1995.

Lenorovitz, D.R., M.D. Phillips, R.S. Ardrey & G.V. Kloster. "A taxonomic approach to characterizing human-computer interaction." In: G. Salvendy (Ed.). *Human-Computer Interaction.* Amsterdam: Elsevier Science Publishers, 111-116, 1984.

Lockwood, A. *Diagram. A visual survey of graphs, maps, charts and diagrams for the graphic designer.* London: Studio Vista, 1969.

Lohse, G., N. Walker, K. Biolsi & H. Rueter. "Classifying graphical information." *Behaviour and Information Technology* 10, 5419-36, 1991.

Luz, S. & Bernsen, N.O. "Interactive advice on the use of speech in multimodal systems design with SMALTO." In: Ostermann, J., K.J. Ray Liu, J.Aa. Sørensen, E. Deprettere & W.B. Kleijn (Eds.). *Proceedings of the Third IEEE Workshop on Multimedia Signal Processing,* Elsinore, Denmark. IEEE, Piscataway, NJ: 489-494, 1999.

Mackinlay, J., S.K. Card & G.G. Robertson. "A semantic analysis of the design space of input devices." *Human-Computer Interaction* 5: 145-90, 1990.

Mullet, K. & D.J. Schiano. "3D or not 3D: 'More is better' or 'Less is more'?" *CHI'95 Conference Companion,* 174-175, 1995.

Rosch, E. "Principles of categorization." In: Rosch, E. & B.B. Lloyd (Eds.). *Cognition and Categorization.* Hillsdale, NJ: Erlbaum, 1978.

SMALTO: http://disc.nis.sdu.dk/smalto/

Stenning, K. & J. Oberlander. "Reasoning with words, pictures and calculi: Computation versus justification." In: Barwise, J., J.M. Gawron, G. Plotkin & S. Tutiya (Eds.). *Situation Theory and Its Applications.* Stanford, CA: CSLI, Vol. 2: 607-62, 1991.

Tufte, E.R. *The Visual Display of Quantitative Information.* Cheshire, CT: Graphics Press, 1983.

Tufte, E.R. *Envisioning information.* Cheshire, CT: Graphics Press, 1990.

Twyman, M. "A schema for the study of graphic language." In: Kolers, P., M. Wrolstad & H. Bouna (Eds.). *Processing of Visual Language* Vol. 1. New York: Plenum Press, 1979.

8. AFFILIATION

Prof. Niels Ole Bernsen
Director of the Natural Interactive Systems Laboratory at the University of Southern Denmark
Main Campus: Odense University
Science Park 10
5230 Odense M, Denmark
URL http://www.nis.sdu.dk

TOM BRØNDSTED, LARS BO LARSEN, MICHAEL MANTHEY, PAUL MC KEVITT, THOMAS B. MOESLUND, AND KRISTIAN G. OLESEN

DEVELOPING INTELLIGENT MULTIMEDIA APPLICATIONS

1. INTRODUCTION

Intelligent multimedia (IntelliMedia), which involves the computer processing and *understanding* of perceptual input from at least speech, text and visual images, and then reacting to it, is complex and involves signal and symbol processing techniques from not just engineering and computer science but also artificial intelligence and cognitive science (Mc Kevitt, 1994, 1995/96, 1997). With IntelliMedia systems, people can interact in spoken dialogues with machines, querying about what is being presented and even their gestures and body language can be interpreted.

Although there for a long time has been much success in developing theories, models and systems in the areas of Natural Language Processing (NLP) and Vision Processing (VP) (Partridge, 1991; Rich & Knight, 1991) there has until recently been little progress in integrating these two sub-areas of Artificial Intelligence (AI). In the beginning, although the general aim of AI was to build integrated language and vision systems, few systems were paying equal attention to both fields. As a result, the two isolated sub-fields of NLP and VP quickly arose. It is not clear why there has not already been much more activity in integrating NLP and VP. Is it because of the long-time reductionist trend in science up until the recent emphasis on chaos theory, non-linear systems, and emergent behaviour? Or, is it because the people who have tended to work on NLP tend to be in other Departments, or of a different ilk, from those who have worked on VP? Dennett (1991, p. 57-58) said "Surely a major source of the widespread scepticism about 'machine understanding' of natural language is that such systems almost never avail themselves of anything like a visual workspace in which to parse or analyse the input. If they did, the sense that they were actually understanding what they processed would be greatly heightened (whether or not it would still be, as some insist, an illusion). As it is, if a computer says, 'I see what you mean' in response to input, there is a strong temptation to dismiss the assertion as an obvious fraud."

People are able to combine the processing of language and vision with apparent ease. In particular, people can use words to describe a picture, and can reproduce a picture from a language description. Moreover, people can exhibit this kind of behaviour over a very wide range of input pictures and language descriptions. Although there are theories of how we process vision and language, there are few theories about how such processing is integrated. There have been large debates in

B. Granström et al. (eds.), Multimodality in Language and Speech Systems, 149–171.

psychology and philosophy with respect to the degree to which people store knowledge as propositions or pictures (Kosslyn & Pomerantz, 1977; Pylyshyn, 1973).

There are at least two advantages of linking the processing of natural languages to the processing of visual scenes. First, investigations into the nature of human cognition may benefit. Such investigations are being conducted in the fields of psychology, cognitive science, and philosophy. Computer implementations of integrated VP and NLP can shed light on how people do it. Second, there are advantages for real-world applications. The combination of two powerful technologies promises new applications: automatic production of speech/text from images; automatic production of images from speech/text, and the automatic interpretation of images with speech/text. The theoretical and practical advantages of linking natural language and vision processing have also been described in Wahlster (1988).

Early work for synthesising simple text from images was conducted by Waltz (1975) who produced an algorithm capable of labelling edges and corners in images of polyhedra. The labelling scheme obeys a constraint minimisation criterion so that only sets of consistent labellings are used. The system can be expected to become 'confused' when presented with an image where two mutually exclusive but self-consistent labellings are possible. This is important because in this respect the program can be regarded as perceiving an illusion such as what humans see in the Necker cube. However, the system seemed to be incapable of any higher-order text descriptions. For example, it did not produce natural language statements such as "There is a cube in the picture."

A number of natural language systems for the description of image sequences have been developed (Herzog & Retz-Schmidt, 1990; Neumann & Novak, 1986). These systems can verbalize the behaviour of human agents in image sequences about football and describe the spatio-temporal properties of the behaviour observed. Retz-Schmidt (1991) and Retz-Schmidt & Tetzlaff (1991) describe an approach which yields plan hypotheses about intentional entities from spatio-temporal information about agents. The results can be verbalized in natural language. The system called REPLAI-II takes observations from image sequences as input. Moving objects from two-dimensional image sequences have been extracted by a vision system (Herzog et al., 1989) and spatio-temporal entities (spatial relations and events) have been recognised by an event-recognition system. A focussing process selects interesting agents to be concentrated on during a plan-recognition process. Plan recognition provides a basis for intention recognition and plan-failure analysis. Each recognised intentional entity is described in natural language. A system called SOCCER (André et al., 1988; Herzog et al., 1989) verbalizes real-world image sequences of soccer games in natural language and REPLAI-II extends the range of capabilities of SOCCER. Here, NLP is used more for annotation through text generation with less focus on analysis.

Maaß et al. (1993) describe a system, called *Vitra Guide*, that generates multimodal route descriptions for computer assisted vehicle navigation. Information is presented in natural language, maps and perspective views. Three classes of spatial relations are described for natural language references: (1) topological relations (e.g. in, near), (2) directional relations (e.g. left, right) and (3) path relations (e.g. along, past). The output for all presentation modes relies on one common 3D model of the domain. Again, Vitra emphasizes annotation through generation of text, rather than analysis, and the vision module considers interrogation of a database of digitised road and city maps rather than vision analysis.

Some of the engineering work in NLP focuses on the exciting idea of incorporating NLP techniques with speech, touch-screen, video and mouse to provide advanced multimedia interfaces (Maybury, 1993; Maybury & Wahlster, 1998). Examples of such work are found in the ALFresco system which is a multimedia interface providing information on Italian Frescoes (Carenini et al., 1992, and Stock, 1991), the WIP system that provides information on assembling, using, and maintaining physical devices like an espresso machine or a lawnmower (André & Rist, 1993, and Wahlster et al., 1993), and a multimedia interface which identifies objects and conveys route plans from a knowledge-based cartographic information system (Maybury, 1991).

Others developing general IntelliMedia platforms *include Situated Artificial Communicators* (Rickheit & Wachsmuth, 1996), *Communicative Humanoids* (Thórisson, 1996, 1997), *AESOPWORLD* (Okada, 1996, 1997), multimodal interfaces like *INTERACT* (Waibel et al., 1996), and *SmartKom* (Wahlster et al., 2001). Other moves towards integration are reported in Denis & Carfantan (1993), Mc Kevitt (1994, 1995/96) and Pentland (1993).

2. CHAMELEON AND THE INTELLIMEDIA WORKBENCH

The Institute for Electronic Systems at Aalborg University, Denmark, has expertise in the area of IntelliMedia and has established an initiative called IntelliMedia 2000+ funded by the Faculty of Science and Technology. IntelliMedia 2000+ involves four research groups from three Departments within the Institute for Electronic Systems: Computer Science (CS), Medical Informatics (MI), Laboratory of Image Analysis (LIA) and Center for PersonKommunikation (CPK), focusing on platforms for integration and learning, expert systems and decision taking, image/-vision processing, and spoken language processing/sound localisation, respectively. The first two groups provide a strong basis for methods of integrating semantics and conducting learning and decision taking while the latter groups focus on the two main input/output components of IntelliMedia, vision and speech/sound. More details on IntelliMedia 2000+ can be found at http://www.cpk.auc.dk/imm.

IntelliMedia 2000+ has developed the first prototype of an IntelliMedia software and hardware platform called CHAMELEON, which is general enough to be used for a number of different applications. CHAMELEON demonstrates that existing software modules for (1) distributed processing and learning, (2) decision taking, (3) image processing, and (4) spoken dialogue processing can be interfaced to a single platform and act as communicating agent modules within it. CHAMELEON is independent of any particular application domain and the various modules can be distributed over different machines. Most of the modules are programmed in C++, C, and Java. More details on CHAMELEON and the IntelliMedia WorkBench can be found in Brøndsted et al. (1998)

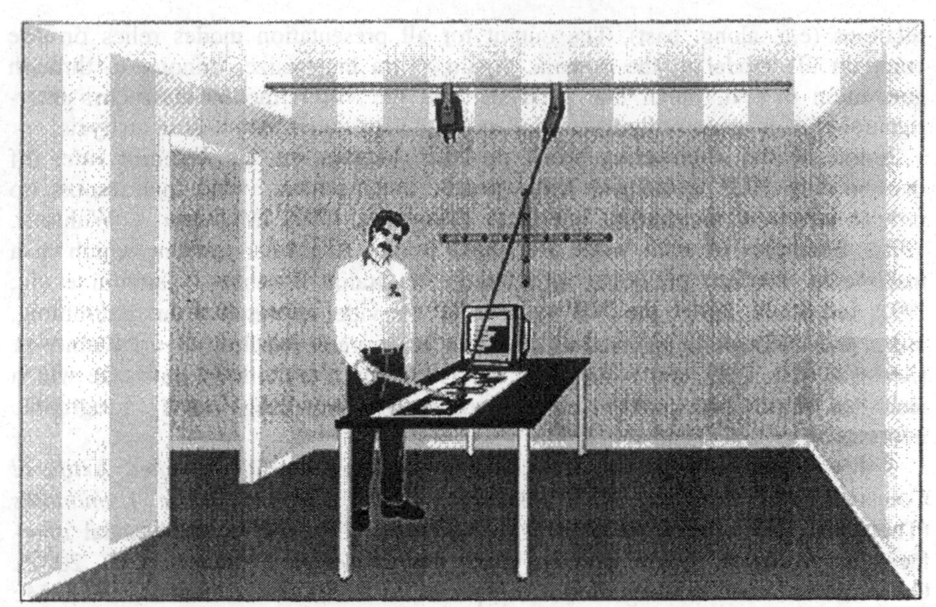

Figure . 1 Physical layout of the IntelliMedia WorkBench.

2.1. IntelliMedia WorkBench

One of the applications of CHAMELEON is the *IntelliMedia WorkBench,* which is a hardware and software platform as shown in Figure 1. One or more cameras and lasers can be mounted in the ceiling, microphone array placed on the wall and there is a table where things (objects, gadgets, people, pictures, 2D/3D models, building plans, etc.) can be placed. The domain of this application is *a Campus Information System,* which gives information on the architectural and functional layout of a building. 2-dimensional (2D) architectural plans of the building drawn on white paper are laid on the table and the user can ask questions about them. The plans represent two floors of the 'A2' building at Fredrik Bajers Vej 7, Aalborg University.

In this setup, there is one static camera which calibrates the plans on the table and the laser, and interprets the user's pointing while the system points to locations and draws routes with a laser. Inputs are simultaneous speech and/or pointing gestures and outputs are synchronised speech synthesis and pointing. In this application, all of CHAMELEON can be run on a single standard Intel computer, which handles input for the Campus Information System in real-time.

The 2D plan, which is placed on the table, is printed out on A0 paper having the dimensions: 84x118cm. Due to the size of the pointer's tip (2x1cm), the size of the table, the resolution of the camera and uncertainty in the tracking algorithm, a size limitation is introduced. The smallest room in the 2D plan, which is a standard office, cannot be less than 3 cm wide. The size of a standard office on the printout is 3x4 cm which is a feasible size for the system. The 2D plan is shown in Figure 2.

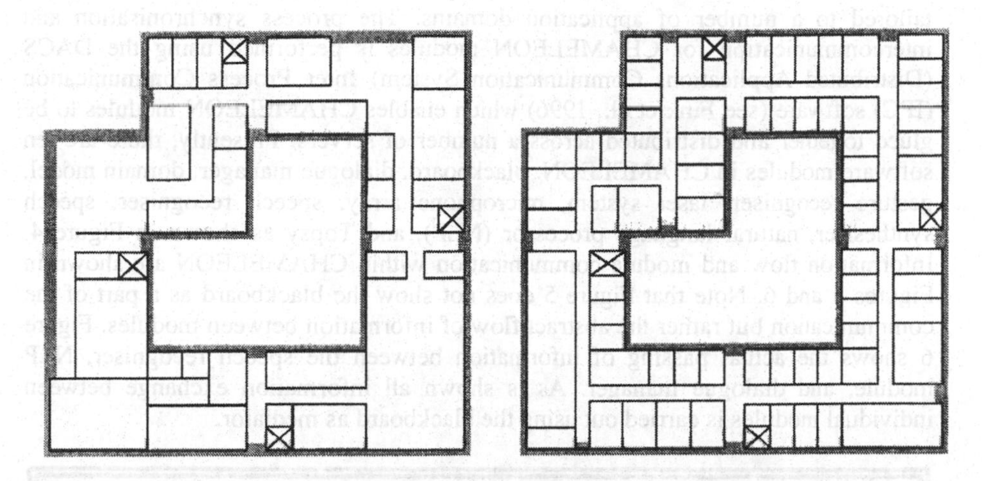

Figure 2. 2D plan of the 'A2' building at Fredrik Bajers Vej 7, Aalborg University. Left: ground floor; Right: 1st floor.

A sample dialogue, which the first prototype can process, is shown in Figure.3. The example includes user intentions, which are instructions and queries, and exophoric/deictic reference.

USER:	Show me Thomas' office.
CHAMELEON:	[points] This is Thomas' office.
USER:	Where is the computer room?
CHAMELEON:	[points] The computer room is here.
USER:	[points to instrument repair] Whose office is this?
CHAMELEON:	[points] This is not an office, this is instrument repair.
USER:	Show me the route from Lars Bo Larsen's office to Hanne Gade's office
CHAMELEON:	[draws route] This is the route from Lars Bo's office to Hanne's office.
. . .	

Figure 3. Simple sample dialogue

2.2. Architecture of CHAMELEON

CHAMELEON has a distributed architecture of communicating agent modules processing inputs and outputs from different modalities and each of which can be tailored to a number of application domains. The process synchronisation and intercommunication for CHAMELEON modules is performed using the DACS (Distributed Applications Communication System) Inter Process Communication (IPC) software (see Fink et al., 1996) which enables CHAMELEON modules to be glued together and distributed across a number of servers. Presently, there are ten software modules in CHAMELEON: blackboard, dialogue manager, domain model, gesture recogniser, laser system, microphone array, speech recogniser, speech synthesiser, natural language processor (NLP), and Topsy as shown in Figure 4. Information flow and module communication within CHAMELEON are shown in Figures 5 and 6. Note that Figure 5 does not show the blackboard as a part of the communication but rather the abstract flow of information between modules. Figure 6 shows the actual passing of information between the speech recogniser, NLP module, and dialogue manager. As is shown all information exchange between individual modules is carried out using the blackboard as mediator.

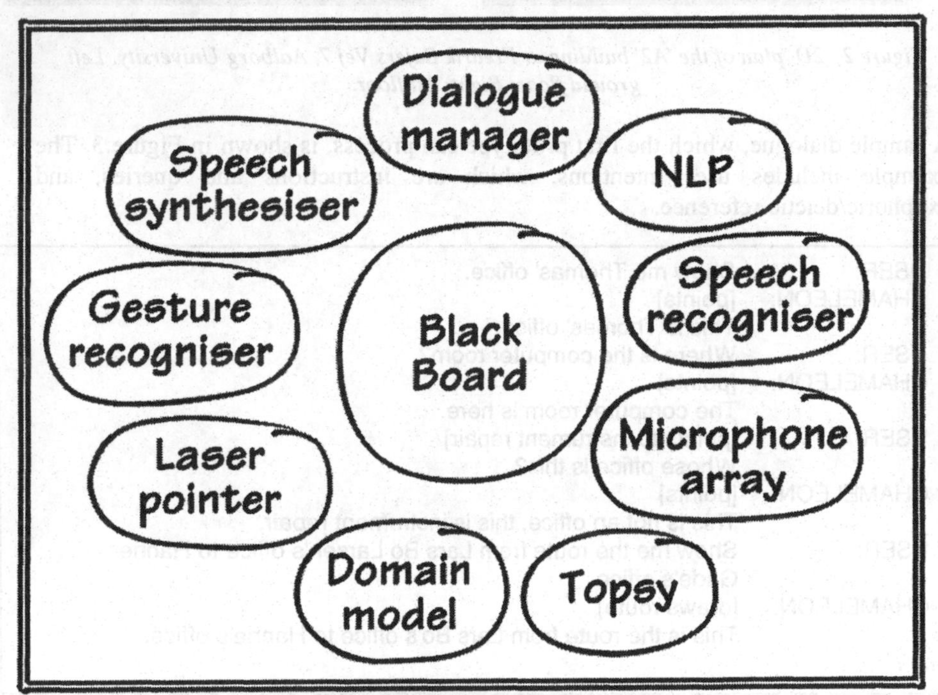

Figure 4. Architecture of CHAMELEON

As the intention is that no direct interaction between modules need take place the architecture is modularised and open but there are possible performance costs. However, nothing prohibits direct communication between two or more modules if this is found to be more convenient. For example, the speech recogniser and NLP modules can interact directly as the parser needs every recognition result anyway and at present no other module has use for output from the speech recogniser. The blackboard and dialogue manager form the kernel of CHAMELEON. We shall now give a brief description of each module.

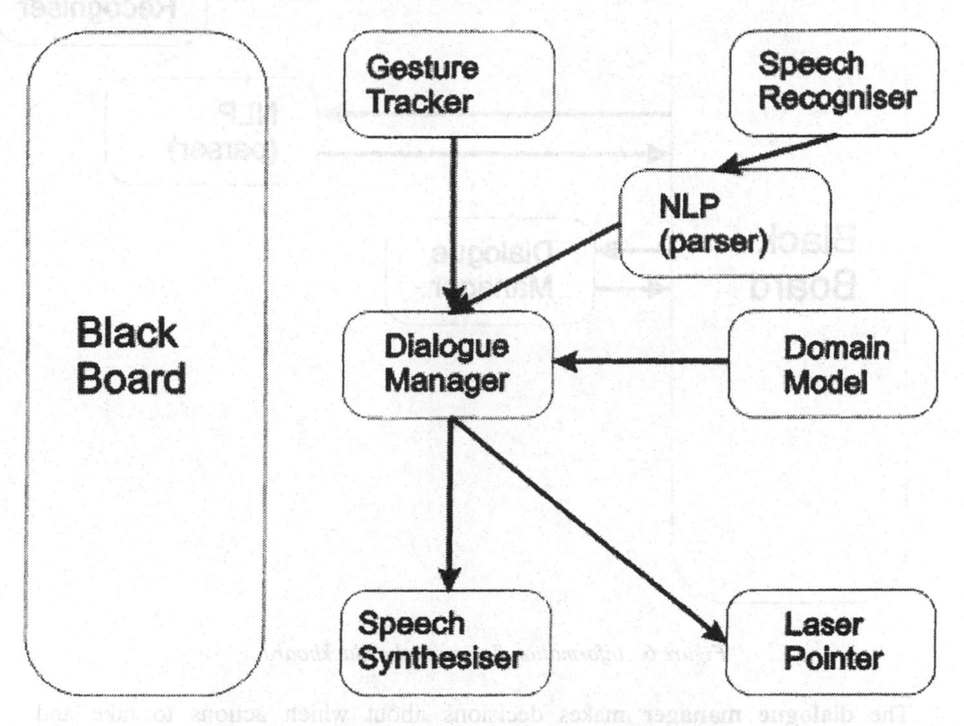

Figure 5. Information flow and module communication

The **blackboard** stores semantic representations produced by each of the other modules and keeps a history of these over the course of an interaction. All modules communicate through the exchange of semantic representations with each other or the blackboard. Semantic representations are frames in the spirit of Minsky (1975) and our frame semantics consists of (1) input, (2) output, and (3) integration frames for representing the meaning of intended user input and system output. The intention is that all modules in the system will produce and read frames. Frames are coded in CHAMELEON as messages built of predicate-argument structures following a specific BNF definition. The frame semantics was presented in Mc Kevitt & Dalsgaard (1997).

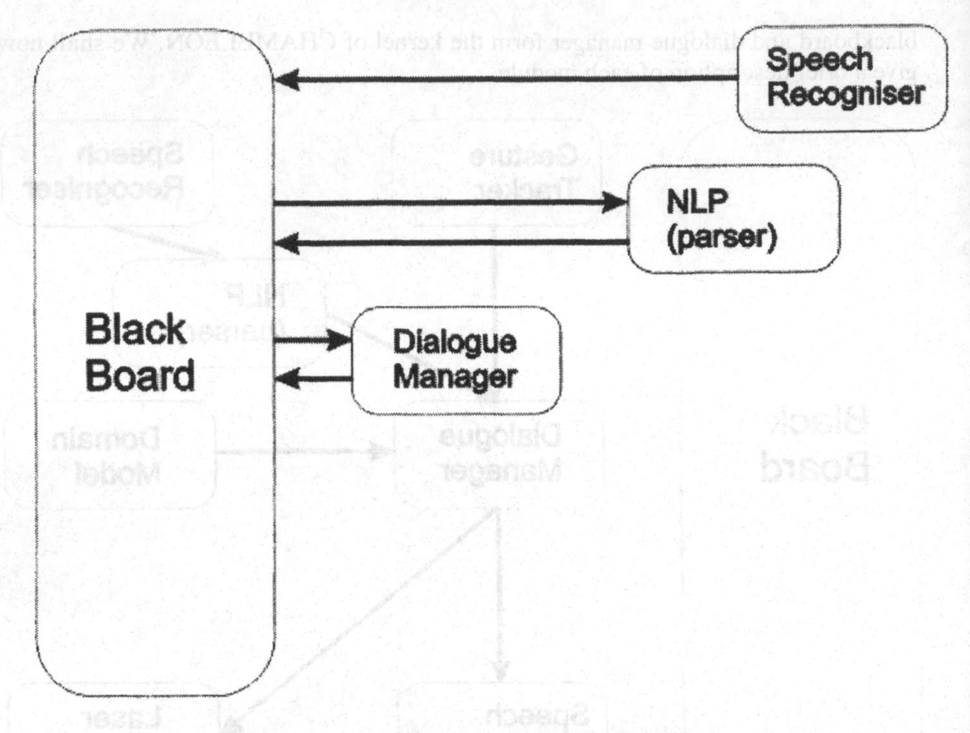

Figure 6. Information flow with the blackboard

The **dialogue manager** makes decisions about which actions to take and accordingly sends commands to the output modules (laser and speech synthesiser) via the blackboard. At present the functionality of the dialogue manager is to integrate and react to information coming in from the speech/NLP and gesture modules and to sending synchronised commands to the laser system and the speech synthesiser modules. Phenomena such as managing clarification sub-dialogues where CHAMELEON has to ask questions are not included at present. It is hoped that in future prototypes the dialogue manager will enact more complex decision taking over semantic representations from the blackboard using, for example, the HUGIN software tool (Jensen F., 1996) based on Bayesian Networks (Jensen F.V., 1996).

The **domain model** contains a database of all locations and their functionality, tenants and coordinates. The model is organised in a hierarchical structure: areas, buildings and rooms. Rooms are described by an identifier for the room (room number) and the type of the room (office, corridor, toilet, etc.). The model includes functions that return information about a room or a person. Possible inputs are coordinates or room number for rooms and name for persons, but in principle any attribute can be used as key and any other attribute can be returned. Furthermore, a path planner is provided, calculating the shortest route between two locations.

A design principle of imposing as few physical constraints as possible on the user (e.g. data gloves or touch screens) leads to the inclusion of a vision based

gesture recogniser. Currently, it tracks a pointer via a camera mounted in the ceiling. Using one camera, the gesture recogniser is able to track 2D pointing gestures in real time. Only two gestures are recognised at present: pointing and not-pointing. Experiments with recognition of other more complex kinds of gestures like marking an area and indicating a direction (with hands and fingers) are reported in (Brøndsted, 1999c).

The camera continuously captures images that are digitised by a frame-grabber. From each digitised image the background is subtracted leaving only the motion (and some noise) within this image. This motion is analysed in order to find the direction of the pointing device and its tip. By temporal segmenting of these two parameters, a clear indication of the position the user is pointing to at a given time is found. The error of the tracker is less than one pixel (through an interpolation process) for the pointer.

A **laser system** acts as a "system pointer". It can be used for pointing to positions, drawing lines and displaying text. The laser beam is controlled in real-time (30 kHz). It can scan frames containing up to 600 points with a refresh rate of 50 Hz thus drawing very steady images on surfaces. It is controlled by a standard Intel PC host computer. The pointer tracker and the laser pointer have been carefully calibrated so that they can work together. An automatic calibration procedure has been set up involving both the camera and laser where they are tested by asking the laser to follow the pointer.

A **microphone array** (Leth-Espensen & Lindberg, 1996) is used to locate sound sources, e.g. a person speaking. Depending upon the placement of a maximum of 12 microphones it calculates sound source positions in 2D or 3D. It is based on measurement of the delays with which a sound wave arrives at the different microphones. From this information the location of the sound source can be identified. Another application of the array is to use it to focus at a specific location thus enhancing any acoustic activity at that location. This module is in the process of being incorporated into CHAMELEON.

Speech recognition is handled by the grapHvite real-time continuous speech recogniser (Power et al., 1997) or any recogniser supporting the HTK format for acoustic models. This includes the latest CPK speech recogniser (Christensen et al., 1998; Olsen, 2000). It is based on HMMs (Hidden Markov Models) of generalised triphones for acoustic decoding of English or Danish. For training of HMMs, the SpeechDat(II) corpus has been employed (Lindberg, 1999). The recognition process focuses on recognition of speech concepts and ignores non-content words or phrases. A finite state network describing phrases can be created by hand in accordance with the domain model and the grammar for the natural language parser. However, the language model can also be generated automatically by a grammar converter in the NLP module (see below). The speech recogniser takes speech signals as input and produces text strings (1-best, n-best lists) as output.

We use the Infovox Text-To-Speech (TTS) **speech synthesiser** that is capable of synthesising Danish and English (Infovox, 1994). It is a rule based formant synthesiser and can simultaneously cope with multiple languages, e.g. pronounce a Danish name within an English utterance. Infovox takes text as input and produces speech as output. Integration of the CPK diphone-based speech synthesiser (Andersen et al., 2000), which is under development for Danish, is being considered.

Natural language processing is based on a compound feature based (so-called unification) grammar formalism for extracting semantics from the one-best utterance text output from the speech recogniser (Brøndsted, 1999a, 1999b). The parser carries out a syntactic constituent analysis of input and subsequently maps values into semantic frames. The rules used for syntactic parsing are based on a subset of the EUROTRA formalism, i.e. in terms of lexical rules and structure building rules (Bech, 1991). Semantic rules define certain syntactic sub-trees and which frames to create if the sub-trees are found in the syntactic parse trees. The module is also capable of generating finite state approximations of the unification grammars to be used by a standard grammar constrained speech recogniser like graphHvite. A natural language generator has been constructed, however, so far generation is conducted by using canned text.

The basis of the Phase Web paradigm (Manthey, 1998), and its incarnation in the form of a program called **Topsy**, is to represent knowledge and behaviour in the form of hierarchical relationships between the mutual exclusion and co-occurrence of events. In AI parlance, Topsy is a distributed, associative, continuous-action, dynamic partial-order planner that learns from experience. Relative to MultiMedia, integrating independent data from multiple media begins with noticing that what ties otherwise independent inputs together is the fact that they occur simultaneously. This is also Topsy's basic operating principle, but this is further combined with the notion of mutual exclusion, and thence to hierarchies of such relationships (Manthey, 1998).

3. FRAME SEMANTICS

As mentioned above, the meaning of interactions over the course of the multimodal dialogue is represented using a frame semantics. All modules in the system should be able to produce and read frames. Frames are coded in CHAMELEON with messages built as predicate-argument structures following a specific BNF definition (see Brøndsted et al., 1998). Frames represent some crucial elements such as *module, input/output, intention, location,* and *timestamp*. Module is simply the name of the module producing the frame (e.g. NLP). Inputs are the input recognised whether spoken (e.g. “Show me Hanne’s office”) or gestures (e.g. pointing coordinates) and outputs the intended output whether spoken (e.g. “This is Hanne's office.”) or gestures (e.g. pointing coordinates). Timestamps can include the times a given module commenced and terminated processing and the time a frame was written on the blackboard. The frame semantics also includes representations for two key phenomena in language/vision integration: reference and spatial relations.

Frames can be grouped into three categories: (1) *input*, (2) *output* and (3) *integration*. Input frames are those, which come from modules processing perceptual input, output frames are those produced by modules generating system output, and integration frames are integrated meaning representations constructed over the course of a dialogue (i.e. all other frames). In Appendix A, we give a more detailed description of the frame semantics of CHAMELEON.

4. DACS

DACS is currently the communication system for CHAMELEON and the IntelliMedia WorkBench and is used to glue all the modules together enabling communication between them. Applications of CHAMELEON typically consist of several interdependent modules, often running on separate machines or even dedicated hardware. This is indeed the case for the IntelliMedia WorkBench application. Such distributed applications have a need to communicate in various ways. Some modules feed others in the sense that all generated output from one is treated further by another. In the Campus Information System all modules report their output to the blackboard where it is stored. Although our intention is currently to direct all communication through the blackboard, we could just as well have chosen to simultaneously transfer output to several modules. For example, utterances collected by the speech recogniser can be sent to the blackboard but also sent simultaneously to the NLP module that may become relevant when efficiency is an important issue.

Another kind of interaction between processes is through remote procedure calls (RPCs), which can be either *synchronous* or *asynchronous*. By synchronous RPCs we understand procedure calls where we want immediate feedback, that is, the caller stops execution and waits for an answer to the call. In the Campus Information System this could be the dialogue manager requesting the last location to which a pointing event occurred. In the asynchronous RPC, we merely submit a request and carry on with any other task. This could be a request to the speech synthesiser to produce an utterance for the user or to the laser to point to some specific location. These kinds of interaction should be available in a uniform way in a heterogeneous environment, without specific concern about what platform the sender and receiver run on.

All these facilities are provided by the Distributed Applications Communication System (DACS) developed at the University of Bielefeld, Germany (Fink et al., 1995, 1996), where it was designed as part of a larger research project developing an IntelliMedia platform (Rickheit & Wachsmuth, 1996) discussed further in the next section. DACS uses a communication demon on each participating machine that runs in user mode, allows multiple users to access the system simultaneously and does not provide a virtual machine dedicated to a single user. The demon acts as a router for all internal traffic and establishes connections to demons on remote machines. Communication is based on simple asynchronous message passing with some extensions to handle dynamic reconfigurations of the system during runtime. DACS also provides on top more advanced communication semantics like RPCs (synchronous and asynchronous) and *demand streams* for handling data parts in continuous data streams. All messages transmitted are recorded in a Network Data Representation that includes type and structure information. Hence, it is possible to inspect messages at any point in the system and to develop generic tools that can handle any kind of data. DACS uses POSIX threads to handle connections independently in parallel. A database in a central name service stores the system configuration to keep the network traffic low during dynamic reconfigurations. A DACS Debugging Tool (DDT) allows inspection of messages before they are delivered, monitoring configurations of the system, and status on connections.

5. RELATION TO OTHER WORK

Situated Artificial Communicators (SFB-360) (Rickheit & Wachsmuth, 1996) is a collaborative research project at the University of Bielefeld, Germany, which focuses on modelling the performance of a person when with he cooperatively with a partner solves a simple assembly task in a given situation. The object chosen is a model airplane (Baufix) to be constructed by a robot from the components of a wooden building kit with instructions from a human. SFB-360 includes equivalents of the modules in CHAMELEON although there is no learning module competitor to Topsy. What SFB-360 gains in size it may loose in integration, i.e. it is not clear yet that all the technology from the subprojects have been fitted together and in particular what exactly the semantic representations passed between the modules are. The DACS process communication system currently used in CHAMELEON is a useful product from SFB-360.

Gandalf is a communicative humanoid which interacts with users in Multimodal dialogue through using and interpreting gestures, facial expressions, body language and spoken dialogue (Thórisson, 1997, 2001). Gandalf is an application of an architecture called *Ymir* which includes perceptual integration of multimodal events, distributed planning and decision making, layered input analysis and motor-control with human-like characteristics and an inherent knowledge of time. Ymir has a blackboard architecture and includes modules equivalent to those in CHAMELEON. However, there is no vision/image processing module in the sense of using cameras since gesture tracking is done with the use of a data glove and body tracking suit and an eye tracker is used for detecting the user's eye gaze. However, it is anticipated that Ymir could easily handle the addition of such a vision module if one were needed. Ymir has no learning module equivalent to Topsy. Ymir's architecture is even more distributed than CHAMELEON's with many more modules interacting with each other. Ymir's semantic representation is much more distributed with smaller chunks of information than our frames being passed between modules.

AESOPWORLD is an integrated comprehension and generation system for integration of vision, language and motion (Okada, 1997). It includes a model of mind consisting of nine domains according to the contents of mental activities and five levels along the process of concept formation. The system simulates the protagonist or fox of an AESOP fable, "the Fox and the Grapes", and his mental and physical behaviour are shown by graphic displays, a voice generator, and a music generator which expresses his emotional states. AESOPWORLD has an agent-based distributed architecture and also uses frames as semantic representations. It has many modules in common with CHAMELEON although again there is no vision input to AESOPWORLD, which uses computer graphics to depict scenes.

AESOPWORLD has an extensive planning module but conducts more traditional planning than CHAMELEON's Topsy.

The INTERACT project (Waibel et al., 1996) involves developing Multimodal Human Computer Interfaces including the modalities of speech, gesture and pointing, eye-gaze, lip motion and facial expression, handwriting, face recognition and tracking, and sound localisation. The main concern is with improving recognition accuracies of modality specific component processors as well as developing optimal combinations of multiple input signals to deduce user intent more reliably in cross-modal speech-acts. INTERACT also uses a frame representation for integrated semantics from gesture and speech and partial hypotheses are developed in terms of partially filled frames. The output of the interpreter is obtained by unifying the information contained in the partial frames. Although Waibel et al. (1996) present good work on multimodal interfaces it is not clear that they have developed an integrated platform, which can be used for developing multimodal applications.

SmartKom (Wahlster et al., 2001) has a stronger focus on general platform issues where the applications (e.g. an on-line cinema ticket reservation system) merely serve the exploration of communication paradigms. Structures comparable with the semantic frames of CHAMELEON are represented in a well-formed XML-based mark-up language and communication between modules is carried out via a blackboard architecture. The central idea of this system is that the user is supposed to "delegate" tasks (e.g. the reservation of a ticket) to a life-like communication agent and if necessary to help the agent carry out this task.

6. CONCLUSION AND FUTURE WORK

We have described the architecture and functionality of CHAMELEON: an open, distributed architecture with ten modules glued into a single platform using the DACS communication system. We described the IntelliMedia WorkBench application, a software and physical platform where a user can ask for information about things on a physical table. The current domain is *a Campus Information System* where 2D building plans are placed on the table and the system provides information about tenants, rooms and routes and can answer questions like "Whose office is this?" in real time. CHAMELEON fulfills the goal of developing a general platform for integration of at least language/vision processing which can be used for research but also for student projects as part of the Master's degree education.

There are a number of avenues for future work with CHAMELEON. We would like to process dialogue that includes examples of (1) spatial relations and (2) anaphoric reference. It is hoped that more complex decision taking can be introduced to operate over semantic representations in the dialogue manager or blackboard using, for example, the HUGIN software tool (Jensen F., 1996). The gesture module will be augmented so that it can handle gestures other than pointing. Topsy will be asked to do more complex learning and processing of input/output from frames. The microphone array has to be integrated into CHAMELEON and set to work. Also, at present CHAMELEON is static and it might be interesting to see how it performs whilst being integrated with a web-based virtual or real robot or as part of an intellimedia videoconferencing system where multiple users can direct cameras through spoken dialogue and gesture. A miniature version of this idea has already been completed as a student project (Bakman et al., 1997).

Intelligent MultiMedia will be important in the future of international computing and media development and IntelliMedia 2000+ at Aalborg University, Denmark, brings together the necessary ingredients from research, teaching and links to industry to enable its successful implementation. The CHAMELEON platform and IntelliMedia WorkBench application are ideal for testing integrated processing of language and vision.

7. APPENDIX

Appendix A: Details of Frame Semantics

Here, we discuss frames with a focus more on frame semantics than on frame syntax. In fact the actual coding of frames as messages within CHAMELEON has a different syntax (for an exact BNF definition, see Brøndsted et al., 1998).

Input frames

An input frame takes the general form:

[MODULE
INPUT: input
INTENTION: intention-type
TIME: timestamp]

where MODULE is the name of the input module producing the frame, INPUT can be at least UTTERANCE or GESTURE, *input* is the utterance or gesture and *intention-type* includes different types of utterances and gestures. An utterance input frame can at least have intention-type (1) query? (2) instruction! and (3) declarative. An example of an utterance input frame is:

[SPEECH-RECOGNISER
UTTERANCE: (Point to Hanne's office)
INTENTION: instruction!
TIME: timestamp]

A gesture input frame is where *intention-type* can be at least (1) pointing, (2) mark-area, and (3) indicate-direction. An example of a gesture input frame is:

[GESTURE
GESTURE: coordinates (3, 2)
INTENTION: pointing
TIME: timestamp]

Output frames

An output frame (F-out) takes the general form:

[MODULE
INTENTION: intention-type
OUTPUT: output
TIME: timestamp]

where MODULE is the name of the output module producing the frame, intention-type includes different types of utterances and gestures and OUTPUT is at least UTTERANCE or GESTURE. An utterance output frame can at least have intention-type (1) query? (2) instruction! and (3) declarative. An example utterance output frame is:

```
[SPEECH-SYNTHESIZER
 INTENTION: declarative
 UTTERANCE: (This is Hanne's office)
 TIME: timestamp]
```

A gesture output frame can at least have intention-type (1) description (pointing), (2) description (route), (3) description (mark-area), and (4) description (indicate-direction). An example gesture output frame is:

```
[LASER
 INTENTION: description (pointing)
 LOCATION: coordinates (5, 2)
 TIME: timestamp]
```

Integration frames

Integration frames are all those other than input/output frames. An example utterance integration frame is:

```
[NLP
 INTENTION: description (pointing)
 LOCATION: office (tenant Hanne) (coordinates (5, 2))
 UTTERANCE: (This is Hanne's office)
 TIME: timestamp]
```

Things become even more complex with the occurrence of references and spatial relationships:

```
[MODULE
 INTENTION: intention-type
 LOCATION: location
 LOCATION: location
 LOCATION: location
 SPACE-RELATION: beside
 REFERENT: person
 LOCATION: location
 TIME: timestamp]
```

An example of such an integration frame is:

[DOMAIN-MODEL
 INTENTION: query? (who)
 LOCATION: office (tenant Hanne) (coordinates (5, 2))
 LOCATION: office (tenant Jorgen) (coordinates (4, 2))
 LOCATION: office (tenant Borge) (coordinates (3, 1))
 SPACE-RELATION: beside
 REFERENT: (person Paul-Dalsgaard)
 LOCATION: office (tenant Paul-Dalsgaard) (coordinates (4, 1))
 TIME: timestamp]

It is possible to derive all the frames produced on a blackboard for example input. Complete blackboard history for the query “"Whose office is this?" + [pointing] (exophoric/deictic reference) is given below. The frames given are placed on the blackboard as they are produced and processed. In these histories we choose to have modules interacting in a completely distributed manner with no single coordinator. The actual current implementation of CHAMELEON has a more top-down coordinating dialogue manager.

USER(G-in,U-in): [points]
 Whose office is this?

PROCESSING(1):
SPEECH-RECOGNISER:
(1) wakes up when it detects registering of U-in
(2) maps U-in into F-in
(3) places and registers F-in on blackboard

FRAME(F-in)(1):
[SPEECH-RECOGNISER
 UTTERANCE: (Whose office is this ?)
 INTENTION: query?
 TIME: timestamp]

PROCESSING(2):
NLP:
(1) wakes up when it detects registering of F-in
(2) maps F-in into F-int
(3) places and registers F-int on blackboard:

FRAME(F-int)(1):
[NLP
 INTENTION: query? (who)
 LOCATION: office (tenant person) (coordinates (X, Y))
 REFERENT: this
 TIME: timestamp]

PROCESSING(3):
DOMAIN-MODEL:
(1) wakes up when it detects registering of F-int
(2) reads F-int and sees it's from NLP
(3) cannot update F-int as does not have a name or coordinates
(4) goes back to sleep

PROCESSING(4):
GESTURE:
(1) wakes up when it detects registering of G-in
(2) maps G-in into F-in
(3) places and registers F-in on blackboard

FRAME(F-in)(2):
[GESTURE
GESTURE: coordinates (3, 2)
INTENTION: pointing
TIME: timestamp]

PROCESSING(5):
DIALOGUE MANAGER:
(1) wakes up when it detects registering of F-in(1) and F-in(2)
(2) reads F-in(1) and F-in(2) and sees they are from SPEECH-RECOGNISER and GESTURE that they have same/close timestamp, there is a query? (with referent) + pointing, in a rhythmic way (synchronized)
(3) dials and fires NLP to read GESTURE

PROCESSING(6):
NLP:
(1) woken up by DIALOGUE-MANAGER and reads F-in(2)
(2) sees F-in(2) is from GESTURE
(3) determines referent of "this" to be (coordinates)
(4) produces updated F-int (coordinates)
(5) places and registers updated F-int on blackboard:

FRAME(F-int)(2):
[NLP
INTENTION: query? (who)
LOCATION: office (tenant person) (coordinates (3, 2))
REFERENT: this
TIME: timestamp]

PROCESSING(7):
DOMAIN-MODEL:
(1) wakes up when it detects registering of F-int
(2) reads F-int and sees it's from NLP
(3) produces updated F-int (tenant)
(4) places and registers updated F-int on blackboard:

FRAME(F-int)(3):
[NLP
 INTENTION: query? (who)
 LOCATION: office (tenant Ipke) (coordinates (3, 2))
 REFERENT: this
 TIME: timestamp]

PROCESSING(8):
NLP:
(1) wakes up when it detects registering of F-int
(2) reads F-int and sees it's from DOMAIN-MODEL
(3) produces updated F-int (intention + utterance)
(4) places and registers updated F-int on blackboard:

FRAME(F-int)(4):
[NLP
 INTENTION: declarative (who)
 LOCATION: office (tenant Ipke) (coordinates (3, 2))
 REFERENT: this
 UTTERANCE: (This is Ipke's office)
 TIME: timestamp]

PROCESSING(9):
LASER:
(1) wakes up when it detects registering of F-int
(2) reads F-int and sees it's from DOMAIN-MODEL
(3) produces F-out (pruning + registering)
(4) places and registers F-out on blackboard:

FRAME(F-out)(1):
[LASER
 INTENTION: description (pointing)
 LOCATION: coordinates (3, 2)
 TIME: timestamp]

PROCESSING(10):
SPEECH-SYNTHESIZER:
(1) wakes up when it detects registering of F-int
(2) reads F-int and sees it's from NLP
(3) produces F-out (pruning + registering) places and registers F-out on blackboard:

FRAME(F-out)(2):
[SPEECH-SYNTHESIZER
INTENTION: description
UTTERANCE: (This is Ipke's office)
TIME: timestamp]

PROCESSING(11):
DIALOGUE-MANAGER:
(1) wakes up when it detects registering of F-out(1) and F-out(2)
(2) reads F-out(1) and F-out(2) and sees they are from LASER and SPEECH-SYNTHESIZER
(3) dials and fires LASER and SPEECH-SYNTHESIZER synchronized
(1) LASER reads G-out and fires G-out
(2) SPEECH-SYNTHESIZER reads U-out and fires U-out

CHAMELEON(G-out): [points]
CHAMELEON(U-out): This is Ipke's office.

8. REFERENCES

Andersen, Ove, C. Hoequist & C. Nielsen. "Danish Research Ministry's Initiative on Text-to-Speech Synthesis". In: *Proceedings of Nordic Signal Processing Symposium*, Kolmården, Sweden, 2000.

André, Elisabeth, G. Herzog & T. Rist. "On the simultaneous interpretation of real-world image sequences and their natural language description: the system SOCCER". In: *Proceedings of the 8th European Conference on Artificial Intelligence* 449-454, Munich, Germany, 1988.

André, Elisabeth & Thomas Rist. "The design of illustrated documents as a planning task". In: *Intelligent multimedia interfaces*. M. Maybury (Ed.), 75-93 Menlo Park, CA: AAAI Press, 1993.

Bakman, Lau, Mads Blidegn, Thomas Dorf Nielsen & Susana Carrasco Gonzalez. *NIVICO - Natural Interface for VIdeo COnferencing*. Project Report (8th Semester), Department of Communication Technology, Institute for Electronic Systems, Aalborg University, Denmark, 1997.

Bech, A. "Description of the EUROTRA framework". In: *The Eurotra Formal Specifications, Studies in Machine Translation and Natural Language Processing*. C. Copeland, J. Durand, S. Krauwer & B. Maegaard (Eds), Vol. 2, 7-40 Luxembourg: Office for Official Publications of the Commission of the European Community, 1991.

Brøndsted, Tom. "The CPK NLP Suite for Spoken Language Understanding." In: *Eurospeech, 6th European Conference on Speech Communication and Technology*, Budapest, 1999a.

Brøndsted, Tom. "The Natural Language Processing Modules in REWARD and IntelliMedia 2000+". In: *LAMBDA 25*, S. Kirchmeier-Andersen, H. Erdman Thomsen (Eds.). Copenhagen Business School, Dep. of Computational Linguistics, 1999b.

Brøndsted, Tom. "Reference Problems in Chameleon". In: *ESCA Tutorial and Research Workshop: Interactive Dialogue in Multi-Modal Systems*. Kloster Irsee, 1999c.

Brøndsted, Tom, P. Dalsgaard, L.B. Larsen, M. Manthey, P. Mc Kevitt, T.B. Moeslund & K.G. Olesen. *A platform for developing Intelligent MultiMedia applications*. Technical Report R-98-1004, Center for PersonKommunikation (CPK), Institute for Electronic Systems (IES), Aalborg University, Denmark, May, 1998.

Carenini, G., F. Pianesi, M. Ponzi & O. Stock. *Natural language generation and hypertext access*. IRST Technical Report 9201-06, Instituto Per La Scientifica E Tecnologica, Loc. Pant e Di Povo, I-138100 Trento, Italy, 1992.

Christensen, Heidi, Borge Lindberg & Pall Steingrimsson. *Functional specification of the CPK Spoken LANGuage recognition research system (SLANG)*. Center for PersonKommunikation, Aalborg University, Denmark, March, 1998.

Denis, M. & M. Carfantan (Eds.). *Images et langages: multimodalite et modelisation cognitive*. Actes du Colloque Interdisciplinaire du Comite National de la Recherche Scientifique, Salle des Conferences, Siege du CNRS, Paris, April, 1993.

Dennett, Daniel. *Consciousness explained*. Harmondsworth: Penguin, 1991.

Fink, G.A., N. Jungclaus, H. Ritter & G. Sagerer. "A communication framework for heterogeneous distributed pattern analysis". In: *Proc. International Conference on Algorithms and Applications for Parallel Processing*. V. L. Narasimhan (Ed.), 881-890 IEEE, Brisbane, Australia, 1995.

Fink, Gernot A., Nils Jungclaus, Franz Kummert, Helge Ritter & Gerhard Sagerer. "A distributed system for integrated speech and image understanding". In: *Proceedings of the International Symposium on Artificial Intelligence*. Rogelio Soto (Ed.), 117-126 Cancun, Mexico, 1996.

Herzog, G., C.-K. Sung, E. André, W. Enkelmann, H.-H. Nagel, T. Rist & W. Wahlster. "Incremental natural language description of dynamic imagery". In: *Wissenbasierte Systeme. 3. Internationaler GI-Kongress*, C. Freksa & W. Brauer (Eds.), 153-162 Berlin: Springer-Verlag, 1989.

Herzog, G. & G. Retz-Schmidt. "Das System SOCCER: Simultane Interpretation und natürlich-sprachliche Beschreibung zeitveranderlicher Szenen". In: *Sport und Informatik*, J. Perl (Ed.), 95-119 Schorndorf: Hofmann, 1990

Infovox. INFOVOX: *Text-to-speech converter user's manual (version 3.4)*. Solna, Sweden: Telia Promotor Infovox AB, 1994.

Jensen, Finn V. *An introduction to Bayesian Networks London*, England: UCL Press, 1996.

Jensen, Frank. "Bayesian belief network technology and the HUGIN system". In: *Proceedings of UNICOM seminar on Intelligent Data Management*. Alex Gammerman (Ed.), 240-248 Chelsea Village, London, England, April, 1996.

Kosslyn, S.M. & J.R. Pomerantz. Imagery, propositions and the form of internal representations. In *Cognitive Psychology*, 9, 52-76, 1977.

Leth-Espensen, P. & B. Lindberg. "Separation of speech signals using eigenfiltering in a dual beamforming system". In: *Proc. IEEE Nordic Signal Processing Symposium (NORSIG)*. Espoo, Finland, September, 235-238, 1996.

Lindberg, Børge. "The Danish SpeechDat(II) Corpus - a Spoken Language Resource". In: *Datalingvistisk Forenings Årsmøde 1999 i København.. Proceedings.* CST Working Papers. Report No. 3, B. Maegaard, C. Povlsen & J. Wedekind (Eds), 1999.

Maaß, Wolfgang, Peter Wizinski & Gerd Herzog. *VITRA GUIDE: Multimodal route descriptions for computer assisted vehicle navigation.* Bereich Nr. 93, Universitat des Saarlandes, FB 14 Informatik IV, Im Stadtwald 15, D-6600, Saarbrucken 11, Germany, February, 1993.

Manthey, Michael J. "The Phase Web Paradigm". In: *International Journal of General Systems, special issue on General Physical Systems Theories.* K. Bowden (Ed.), 1998.

Maybury, Mark. "Planning multimedia explanations using communicative acts". In: *Proceedings of the Ninth American National Conference on Artificial Intelligence (AAAI-91)*, Anaheim, CA, July 14-19, 1991.

Maybury, Mark (Ed.). *Intelligent multimedia interfaces.* Menlo Park, CA: AAAI Press, 1993.

Maybury, Mark & Wolfgang Wahlster (Eds.). *Readings in intelligent user interfaces.* Los Altos, CA: Morgan Kaufmann Publishers, 1998.

Mc Kevitt, Paul. "Visions for language". In: *Proceedings of the Workshop on Integration of Natural Language and Vision processing.* Twelfth American National Conference on Artificial Intelligence (AAAI-94), Seattle, Washington, USA, August, 47-57, 1994.

Mc Kevitt, Paul (Ed.). *Integration of Natural Language and Vision Processing (Vols. I-IV).* Dordrecht, The Netherlands: Kluwer-Academic Publishers, 1995/1996.

Mc Kevitt, Paul. "SuperinformationhighwayS". In: *Sprog og Multimedier.* Tom Brøndsted & Inger Lytje (Eds.), 166-183, Aalborg, Denmark: Aalborg University Press, April, 1997.

Mc Kevitt, Paul & Paul Dalsgaard. "A frame semantics for an IntelliMedia TourGuide". In: *Proceedings of the Eighth Ireland Conference on Artificial Intelligence (AI-97)*, Volume 1 104-111, University of Uster, Magee College, Derry, Northern Ireland, September, 1997.

Minsky, Marvin. "A framework for representing knowledge". In: *The Psychology of Computer Vision.* P.H. Winston (Ed.), 211-217 New York: McGraw-Hill, 1975.

Neumann, B. & H.-J. Novak. "NAOS: Ein System zur natürlichsprachlichen Beschreibung zeitveränderlicher Szenen". In: *Informatik. Forschung und Entwicklung*, 1(1): 83-92, 1986.

Okada, Naoyuki. "Integrating vision, motion and language through mind". In: *Integration of Natural Language and Vision Processing, Volume IV, Recent Advances.* Mc Kevitt, Paul (Ed.), 55-80 Dordrecht, The Netherlands: Kluwer Academic Publishers, 1996.

Okada, Naoyuki. "Integrating vision, motion and language through mind". In: *Proceedings of the Eighth Ireland Conference on Artificial Intelligence (AI-97)*, Volume 1, 7-16 University of Uster, Magee, Derry, Northern Ireland, September, 1997.

Olsen, Jesper. *The SLANG Platform: Design and Philosophy, v. 1.* Technical Report, Center for Person-Kommunikation, Aalborg University, September, 2000.

Partridge, Derek. *A new guide to Artificial Intelligence Norwood*, New Jersey: Ablex Publishing Corporation, 1991.

Pentland, Alex (Ed.). *Looking at people: recognition and interpretation of human action.* IJCAI-93 Workshop (W28) at The 13th International Conference on Artificial Intelligence (IJCAI-93), Chambery, France, August, 1993.

Power, Kevin, Caroline Matheson, Dave Ollason & Rachel Morton. *The grapHvite book (version 1.0)*, Cambridge, England: Entropic Cambridge Research Laboratory Ltd., 1997.

Pylyshyn, Zenon. "What the mind's eye tells the mind's brain: a critique of mental imagery". In: *Psychological Bulletin*, 80, 1-24, 1973.

Rich, Elaine & Kevin Knight. *Artificial Intelligence.* New York: McGraw-Hill, 1991.

Rickheit, Gert & Ipke Wachsmuth. "Collaborative Research Centre 'Situated Artificial Communicators' at the University of Bielefeld, Germany". In: *Integration of Natural Language and Vision Processing, Volume IV, Recent Advances.* Mc Kevitt, Paul (Ed.), 11-16, Dordrecht, The Netherlands: Kluwer Academic Publishers, 1996.

Retz-Schmidt, Gudala. "Recognizing intentions, interactions, and causes of plan failures". In: *User Modelling and User-Adapted Interaction* 1: 173-202, 1991.

Retz-Schmidt, Gudala & Markus Tetzlaff. *Methods for the intentional description of image sequences*. Bereich Nr. 80, Universitat des Saarlandes, FB 14 Informatik IV, Im Stadtwald 15, D-6600, Saarbrucken 11, Germany, August, 1991.

Stock, Oliviero. "Natural language and exploration of an information space: the ALFresco Interactive system". In: *Proceedings of the 12th International Joint Conference on Artificial Intelligence* (IJCAI-91) 972-978, Darling Harbour, Sydney, Australia, August, 1991.

Thórinsson, Kris R. *Communicative humanoids: a computational model of psychosocial dialogue skills*. Ph.D. thesis, Massachusetts Institute of Technology, 1996.

Thórisson, Kris R. "Layered action control in communicative humanoids". In: *Proceedings of Computer Graphics Europe* '97 June 5-7, Geneva, Switzerland, 1997.

Thórisson, Kris R. This book, 2001.

Wahlster, Wolfgang. *One word says more than a thousand pictures: On the automatic verbalization of the results of image sequence analysis*. Bereich Nr. 25, Universitat des Saarlandes, FB 14 Informatik IV, Im Stadtwald 15, D-6600, Saarbrucken 11, Germany, February, 1988.

Wahlster, Wolfgang, Elisabeth André, Wolfgang Finkler, Hans-Jurgen Profitlich & Thomas Rist. "Plan-based integration of natural language and graphics generation". In: *Artificial Intelligence, Special issue on natural language generation*, 63, 387-427, 1993.

Wahlster, Wolfgang, Norbert Reithinger & Anselm Blocher. "SmartKom: Multimodal Communication with a Life-Like Character". In: *Eurospeech, 7th European Conference on Speech Communication and Technology*, Aalborg, 2001.

Waibel, Alex, Minh Tue Vo, Paul Duchnowski & Stefan Manke. "Multimodal interfaces. In: *Integration of Natural Language and Vision Processing, Volume IV, Recent Advances*, Mc Kevitt, Paul (Ed.), 145-165, Dordrecht, The Netherlands: Kluwer Academic Publishers, 1996.

Waltz, David. "Understanding line drawings of scenes with shadows". In: *The psychology of computer vision*, Winston, P.H. (Ed.), 19-91 New York: McGraw-Hill, 1975.

9. AFFILIATIONS

T. Brøndsted, L.B. Larsen, M. Manthey, P. Mc Kevitt, T.B. Moeslund, and Kristian G. Olesen

Institute for Electronic systems (IES), Fredrik Bajers Vej 7, Aalborg University, DK-9220, Aalborg, Denmark,

P. Mc Kevitt, Chair in Intelligent MultiMedia at The University of Ulster (Magee), Derry, Northern Ireland

KRISTINN R. THÓRISSON

NATURAL TURN-TAKING NEEDS NO MANUAL: COMPUTATIONAL THEORY AND MODEL, FROM PERCEPTION TO ACTION

1. INTRODUCTION

t-minus-460 msec

Beth and Alan are sitting at a Fifth Avenue outdoors restaurant in Manhattan. Alan is telling Beth an exciting story about his vacation in Nice. Alan presents the story through gesture and speech. Then Beth's arm starts moving and her neck stiffens.

We, the viewers, know that she's surprised to see an elephant in the middle of Manhattan, and that in 460 milliseconds her arm and hand motion will turn into a well-defined deictic gesture, her eyebrows will rise, and her mouth will open with surprise, at which point Alan will most certainly recognize the signs and look over at the elephant. But right now, at t-minus-460 milliseconds, Beth's gesture is barely recognizable as a communicative action, so Alan doesn't know for sure. And thus, before that all happens, in the next 460 milliseconds, Alan has to decide what to do about Beth's behavior. Should he stop telling his story? Or should he go on, in case Beth is simply adjusting her jacket?

Decisions like these are made by dialogue participants as often as 2-3 times per second. For a 30 minute conversation that's over 5000 decisions. And that's just a fraction of what goes on. How do we do it? Face-to-face dialogue consists of interaction between several complex, dynamic systems — visual and auditory display of information, internal processing, knee-jerk reactions, thought-out rhetoric, learned patterns, social convention, etc. One could postulate that the power of dialogue is a direct result of this fact. However, combining a multitude of systems in one place does not guarantee a coherent outcome such as goal-directed dialogue. For this to happen the systems need to be architected in a way that guides their interaction and ensures that — complex as it may be — the interaction tends towards homeostasis in light of errors and uncertainties, towards the set of goals shared by participants.

Past research into what kinds of architectures might guide such systems has resulted in a broad range of studies that combines linguistics, psychology and artificial intelligence (cf. Maes 1990a, Adler 1989, Grice 1989, Grosz & Sidner 1986, Goodwin 1981). The body of work is impressive. However, much of the work focusing on human behavior and cognition has been descriptive, and not well suited, except in very general ways, for working implementations of artificial systems that can participate in interactive face-to-face dialogue. The divide-and-conquer approach of aca-

B. Granström et al. (eds.), Multimodality in Language and Speech Systems, 173–207.

demic research has further resulted in neglect of real-time interpretation and generation of multimodal behavior, a critical component to any such system. Building a generative model involves identifying the contributing processes and formalizing the interaction of these in a system capable of taking the role of a simulated human participant. To be certain that the model performs to spec, the best way to test it is in actual interaction with humans; without real-time constraints and the complexities of the real world the system could easily fail to address fundamental constraints in human communication, chief among them the march of a real-world clock. For this the system needs both real-time perception and action.

This chapter presents a computational model of natural turn-taking in goal-oriented, face-to-face dialogue. The model demonstrates fluent psycho-social dialogue skills in real-time interactions with human users, perceiving their multimodal actions — speech, prosody, body language, manual gesture, gaze — and generating multimodal behavior as output, including speech, facial expressions, manual gesture, spatial attention via head and eye movements, as well as manipulations in a topic domain. The first half of this chapter presents the theory of the model, formulated as a series of hypotheses. Here we look at prior research, identify the missing pieces and relate this to our computational perspective, rooted in classical and behavior-based artificial intelligence. The theory is not an analysis of the 'turn-taking rules' observed in human dialogue — which vary between cultures and can be separated out (our implementation leans on research in this area) — the theory and model present a *turn-taking mechanism*. The second half describes a model based on these hypotheses, and its implementation. The implementation has been tested with a wide range of users and shows significant promise as a first step in bridging semantic analysis, situated dialogue, discourse structure, auditory perception, computer vision, and action selection under a unifying framework. Performance examples of the prototype are given at the end of the chapter.

Pragmatically speaking, a generative model of turn-taking has the potential to free users from the "vending machine" symptoms that have plagued many communicative computer systems in the past: Arbitrary pauses, beeps, button pushes, and instruction guidelines. Any decent implementation of a generative, multimodal turn-taking model should allow for interaction with machines in the same way human interaction works, supporting seamless, finely-timed turn-taking, giving invisible support to the task and the situated natural language communication at hand — without the need for a manual.

The model presented here assumes no artificial protocols. It builds on work from psychology (Sacks et al. 1974, Duncan 1972) and artificial intelligence (Maes 1990b, Selfridge 1959), and is based on the Ymir mind-model for communicative creatures and humanoids (Thórisson 1996, 1999). The part of this model concerned with turn-taking is called the *Ymir turn-taking model*, YTTM, and it address the full perception-action loop of real-time turn-taking, from (1) the basics of multimodal perception to (2) knowledge representation, (3) decision making, and (4) action generation for gaze, gesture, facial expressions and speech planning and execution. The model assumes a task-oriented dialogue, and interfaces with knowledge systems via a limited set of primitives. It does not address the specifics of topic knowledge, and is therefore complementary to models such as Grozs & Sidner's (1986) focus space model, and Clark and Schaefer's contribution model (Cahn & Brennan 1999, Clark

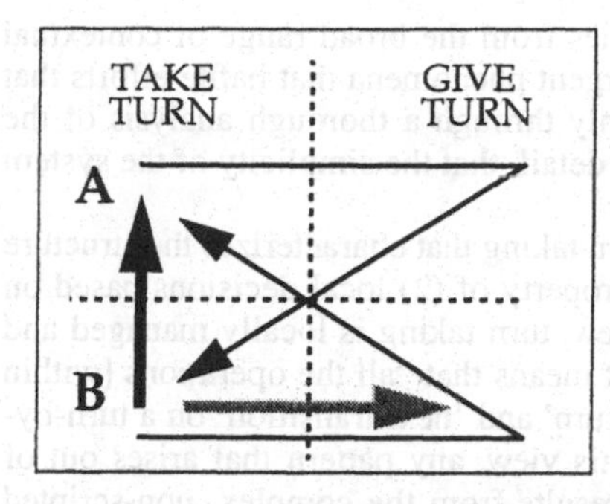

Figure 1. The task of efficient turn transitions includes detecting acceptable transition points. In this figure, Alan (A) and Beth (B) are engaged in dialogue with each other. Beth is talking, Alan listening. Thin arrows demonstrate smooth turns, lower arrow indicates Beth giving turn to Alan; solid bold arrow constitutes an interruption (of B by A) with the possibility of overlapping speech, gray bold arrow shows a failure of the listener, A, to take the turn when it is given (by B), possibly with an unwanted silence.

& Schaefer 1989), which model tracking of dialogue topic in general. YTTM has been implemented in two working systems (Bryson & Thórisson in press, Thórisson 1996), and tested in interaction with human users (Cassell & Thórisson 1999, Thórisson 1996). Results show the broad strokes of dialogue behavior it produces to be acceptable, both in perceiving/generating correct and acceptable turn transitions and in producing the necessary and sufficient turn-taking behaviors in real time, resulting in dialogue patterns similar to that observed in human-human conversation. (We present example interactions in Section 5.4.2.) The encouraging results have prompted the decision to summarize the model's background assumptions, as presented in this chapter. The implementation presented here is limited (1) to two participants, and (2) by the assumption of a single topic. Solutions to both limitations are well understood within the framework of the Ymir architecture (Thórisson 1999), and are currently being worked out.

We will now look at the main features of real-time dialogue as presented in prior research — the features which the model needs to address — and then go on to describe how they relate to the YTTM.

2. PRIOR RESEARCH

When people communicate in face-to-face interaction they take turns speaking (Duncan 1972). The system's main function is to sequentialize information exchange between two or more communicating parties and ensure efficient transmission (Figure 1). It is the key organizing principle of real-time dialogue. The information exchanged during typical face-to-face interaction is constructed through speech, hand gestures, body language, gaze, facial expressions, and multiple combination thereof (Sacks 1992, McNeill 1992, Goodwin 1981). Turn-taking and back-channel feedback (Yngve 1970) have both been shown to be important for conducting successful dialogue (Sacks et al. 1974, Nespolous & Lecours 1986). Turn-taking is, for example, crucial in both negotiation and clarification (Whittaker et al. 1991, Whittaker & Stenton 1988, Sacks et al. 1974).

Goodwin (1981, p. 2) says about the turn:

> "In the abstract, the phenomenon of turn-taking seems quite easy to define. The talk of one party bounded by the talk of others constitutes a turn, with turn-taking being the process through which the party doing the talk of the moment is changed."

Like many before (and after) him he goes on to say that on closer inspection things are not as simple as they look in the abstract. This is certainly true. However, we argue that the complexity in turn-taking comes from the broad range of contextual influences on the processes, resulting in emergent phenomena that baffle efforts that only look at the surface phenomena. It is only through a thorough analysis of the underlying mechanisms, at multiple levels of detail, that the simplicity of the system becomes apparent.

Sacks et al. (1974) put forth a model of turn taking that characterizes the structure of human conversation as (1) an emergent property of (2) local decisions based on (3) prediction by the participants. In their view, turn taking is locally managed and participant-administrated. Local management means that "all the operations [within the system] are 'local', i.e. directed to 'next turn' and 'next transition' on a turn-by-turn basis" (Sacks et al. 1974, p. 725). In this view, any pattern that arises out of interaction is emergent in the sense that it results from the complex, non-scripted interaction between decisions that are made by each conversant with incomplete knowledge and an independent set of interaction rules. They say further (p. 725-6) that

> "...the turn-taking system is a local management system ... in the sense that it operates in such a way as to allow turn-size and turn-order to vary and be under local management, across variations in other parameters, while still achieving both the aim of all turn-taking systems—the organization of 'n at a time'—and the aim of all turn-taking organizations for speech-exchange systems — 'one at a time while speaker change recurs'".

"Party-administration" refers to the fact that the rules of turn-taking are subject to the conversants' control, i.e. that the rules are designed for being used by each participant individually to manage their communication with others. By hypothesizing the existence of turn-constructional units, Sacks et al. were able to model turn taking with only five — albeit relatively complex — rules. But the most important part of their theory is the set of turn-constructional units they propose, which are *sentential*, *clausal*, *phrasal* and *lexical*. (More unit candidates would clearly have resulted if they had included multiple communication modes in their analysis.) According to their theory, these units are used by speakers to construct a turn (i.e. determine transitions). For example, recognizing that a particular sentence of type *S* is being uttered by a speaker, an interpreter can use her knowledge about sentence type *S* to predict when it ends, making it possible to take turns with no gaps. However, Sacks et al. do not specify what kinds of turn-constructional units distinguish one type of utterance — or multimodal act for that matter — from another. If we assume that a listener is continuously looking for clues to classify each utterance we might conclude that the only features that matter are present in the stream of the audio signal. But this would be a mistake: Anyone who ignored all but the audio signal in a multimodal interaction would be throwing away a wealth of information that can be gleaned from the utterer's behavior pertaining to both the content and the process of the dialogue. We can be pretty certain that the 'evidence' people use to classify turn segments includes a number of sources, all the way from gaze to facial gesture to body stance (Taylor & Cameron 1987, Goodwin 1981).

This leaves us with two problems. From a descriptive point of view, the idea of turn-constructional units may be valid, but it says nothing about the way people actually recognize these units. Furthermore, even when a unit is recognized, its length would not be completely predictable, and the task of prediction becomes clearly also a perceptual task: paying attention to cues that signify the end of the unit.[1] The second problem is that the turn-constructional units that Sacks et al. propose are purely based on the audio stream produced. The mistake is to think only of how utterances relate to the turn, when we really need a theory of how *communicative acts* can be constructed in turns.

What is needed is a mechanism that allows sentential, clausal, phrasal and lexical features — as well as all other types of speaker behaviors indicative of dialogue state — to be recognized in real-time and integrated with a discourse participant's actions. Furthermore, Sacks et al.'s (1974) model does not take into account the internal state of cognitive processing of the participants, which clearly also affects the way they respond to cues in the dialogue. Whether we call this yet another class of turn-constructional units or not is beside the point: We will lump all of these ideas together into a bag called *context*. In the section on the YTTM architecture we will present an approach for doing this based on a version of the blackboard architecture (Nii 1989, Selfridge 1959). In the process we will define operationally what 'context' means in this context.

In what seems to be an incompatible approach to that of Sacks' et al., Duncan (1972) proposed the existence of "cues" for turn signalling. Such cues are generated by interlocutors for the purpose of "signaling" to each other the state of the dialogue, such as whether they want the other to take the turn, whether they want to keep the turn, etc. The claim here is that Duncan's cues are simply the features missing from Sacks et al.'s model — the features that conversants use to identify the turn-constructural units, and their boundaries. These, naturally, vary between cultures and individuals — which is why we find it more difficult to interact smoothly with strangers than with people we know. What, exactly, the set of such 'cues' consist of is not easy to determine, and is bound to vary on an individual basis. The best we might do is to create a collection of what may *look* like 'typical' patterns (cues) for a given group of individuals, families, or cultures. A more important first step though, is to identify which kinds of data and channels carry information relevant to dialogue, and to propose mental mechanisms that might be at work for producing — and especially perceiving — such information. This will be our focus in section 4., "Principles of the YTTM".

2.1. *Back-Channel Feedback*

No treatise on turn-taking is complete without a discussion of back-channel feedback (Yngve 1970). Face-to-face interaction quickly breaks down if communication can only happen at or above the turn level (Nespolous & Lecours 1986) — there needs to be a two-way incremental exchange of information within the turn. Part of the task for a listener is to make sure that the other party knows that she is paying attention, and indicate that she is at the same state in the conversation. This is done mainly in the back channel (Yngve 1970). Back channel feedback is in effect information exchange that supports the interaction itself and helps move it along (McNeill 1992, Goodwin 1981). In English speaking countries it includes using

1. Dead-reckoning — the act of committing to a course of action ahead of time and then blindly executing it — would be another way to solve this problem. More on this in section 3.1, "Achieving Seamlessness Through Perceptual Anticipation".

paraverbals such as "m-hm," "aha," etc., indicating confusion, expressing feelings at given points (by facial gesture, laughter, etc.), and indicating attentional focus. The absence of such regulatory gestures from a listener may disrupt the discourse (Dahan, as referenced in Nespolous & Lecours 1986).[2]

While it may rightfully be argued that overlapping talk in the main communication channel is counter-productive because it interferes with the flow of a conversation (Sacks 1992), co-occurring speech in the paraverbal channel does not (Yngve 1970), unless it is misclassified (by the speaker) as being part of the class of accepted turn-transition cues. One rule of thumb definition of back-channel feedback then is that it is the ongoing (communicative) behavior of a dialogue participant that does not change who is in control of the dialogue at the moment. So, whether something "is" back-channel feedback is not based on what an act looks like (morphology) or who has the turn, because what may be intended as back-channel feedback may turn into an interruption if the speaker misinterprets it. For the perceiver of such behavior this is therefore an issue of ongoing *functional classification*. Functional classification is executed continuously by all participants during dialogue. We will discuss functional classification in section 4.4.1, "Functional Analysis: Characterizing the Broad Strokes 'First'".

Back-channel feedback is modeled in the YTTM as resulting from two very different sources: The processing of the *content* of dialogue, e.g. smiling when we find funny the content of what the speaker is saying, or from the mental machinery *orchestrating* the dialogue, e.g. when we look at the speaker to show we are paying attention. We will look at this claim in section 4.3, "Separating Interaction Control from Content Generation & Delivery".

2.2. *Embodiment*

At least two types of spatial constraints are critical to situated conversation: the *location* and *orientation* of conversants to each other and surroundings, referred to here as *positional elements* and *directional elements*, respectively. The position of conversational participants has implications for spatial reference: glances, pointing gestures and direction-giving head nods will be done differently (varying morphology) depending on where the speaker and listener are positioned in space. The display of visual cues such as facial gesture is bound to a specific location, i.e. the participants' faces. A number of turn-taking signals rely on participant location and facial cues (Duncan 1972), and back-channel feedback is often given through the face (Goodwin 1981). Manual gesture is usually done in the area right in front of the gesturer's body (McNeill 1992), and a perceiver needs to be able to locate these in space. Gaze is often used to reference this space (Goodwin 1986), and can be indicative of the kind of gesture being made (McNeill 1992, Goodwin 1981); gesturers tend to look at their own iconic gestures.

Directional elements have to do with how the participants are turned relative to each other, how various body parts are oriented, and how this changes over the

2. Nespolous & Lecours (1986, page 61) say: "... Dahan (see ref., op. cit.) convincingly demonstrated that the absence of regulatory gestures in the behavior of the listener could lead the speaker to interrupt his speech or to produce incoherent discourse."

course of the interaction. When talking face-to-face, most people prefer to orient their bodies approximately 90° to each other (Sommer 1959). Turning your head away right after your partner finishes speaking can indicate that you think he's done and that you are now preparing a response (Goodwin 1981, 1986). All these features require spatial computation of both participants. We will look at how some of these are implemented in section 5.3.2, "Multimodal Integrators (Table 4)".

3. THE LEAP TO GENERATION

From the discussion so far it is clear that a step-lock "transmitter/receiver" model will not be sufficient when imparting multimodal interaction to the computer. Back-channel feedback, interruptions, real-time construction, unforeseen events all hint at a much more complex, dynamic system in which multiple states and events serve to provide a rich context for the participants' mental processing.

As numerous researchers have shown (Walker & Whittaker 1990, Goodwin 1986, Sacks et al. 1974), turn-taking defines the two main roles of conversants, often referred to as 'speaker' and 'listener'. These terms are too limiting to describe the roles of dialogue participants. We will use the terms *content presenter* and *content interpreter* to refer to these roles, respectively. Firstly, this separates the roles of communicating parties from the modes they use for the communication. Secondly, it avoids confusion between turn mechanisms and the act of speaking (when giving back-channel feedback "listeners" can speak without taking the turn). The relation between content presentation and turns is that, generally speaking, one needs to have the turn to present content. In section 4.2, "Presentation and Interpretation: Role-Based Processing", we will explore how each role calls for its own cognition repertoire.

The model of turn-taking advanced by Sacks et al. (1974) is a good descriptive model of turn-taking. A generative model has to go beyond describing surface phenomena in dialogue, however, it has to re-create the surface events observed through a performance model. Given the amount of sensory data and motor control needed for this to happen, the challenge in turn-taking is how to make context-sensitive mechanisms without having to connect everything to everything else. Modularization of the computational processes must be a significant part of a successful model.

The hypotheses on which the Ymir Turn-Taking Model is based create a necessary bridge between a backdrop of relatively coarse-grain studies of turn-taking and dialogue from the psychological and linguistic literature on the one hand, and, on the other, a computational architecture that dictates mental functioning at much smaller levels of granularity. Although the work described can be classified solely under the rubric of artificial intelligence it is inspired by cognitive models, and the following hypotheses are provided to enable future assessment of the model's psychological plausibility. (This task remains outside of our scope here.) The hypotheses represent a theoretic-complete foundation for the creation of YTTM and stand as a generalization of the computational implementation presented in the second half of the chapter.

3.1. *Achieving Seamlessness Through Perceptual Anticipation*

People do not particularly notice the mechanisms by which they take turns speaking. They do not have to pay much attention to how they interweave glances, content, gestures, body movements, etc., seamlessly during conversation. How is this possible? It might be argued that after years of participating in dialogue almost every day, people achieve turn-taking using dead-reckoning.[3] Perhaps they committ to taking the turn several hundreds of milliseconds in advance, and then blindly stick to it, from that moment on running ballistic. This is a valid hypothesis that deserves consideration.

In the context of face-to-face dialogue, which can go on for hours, 100 msec is not a very long time to be spending between turns. Yet, as Goodwin (1981) and other have shown, turn transitions of 100 msec or less, even ones with no pauses in the speech channel, happen frequently in spoken dialogue. Given the complexity of turn-taking, dead-reckoning a long time ahead would greatly increase the likelihood of erroneous turn transitions. If it exists, it is therefore likely to span only a short interval. Of course both parties in a dialogue have a choice reaction time of 100 msec — a speaker can decide to continue an utterance on a whim, destroying any conclusions about a valid turn-transition that the other party may have predicted equally many milliseconds ago. So even for a relatively short turn, lasting, say, 3-4 seconds, a valid turn transition predicted by the interpreter 400-500 msec in advance can be destroyed 200 msec later by the presenter's decision to continue speaking at the end of that segment, leading unavoidably to overlapping speech. (In any goal-directed conversation overlapping speech is considered non-cooperative (Grice 1989) and is thus to be avoided.) An interpreter who has predicted a turn-transition by dead-reckoning will thus also have to monitor, during the exact moment of turn transition, whether this prediction was erroneous. Given the speed of simple reaction, and the price of erroneous dead-reckoning resulting in unwanted speech overlaps and pauses, it is unlikely that any dead-reckoning turn-taking scheme would span longer than 100-200 msec into the future. Moreover, any such dead-reckoning behavior would become useless in interaction with a non-native speaker with a different rhythm, syntax structure and intonation. Whichever way we look at it, no matter whether some amount of dead-reckoning is happening or not in native-speaker turn-taking, we still end up with the conclusion that there has to be ongoing perceptual monitoring during transitions. Moreover, because of the unpredictability of turn-taking, the extent of dead-reckoning is likely to be very short, possibly close to being negligible. The assumption here is that the role of open-loop — ballistic — action in turn-taking is likely to be very small, and can for all practical purposes be ignored.

Both Sacks et al.'s (1974) and Duncan's (1972) work provide evidence that the difficulty of modeling turn-taking lies first and foremost in perception, because a participant has to infer what constitutes a valid turn-giving "signal" solely from perceptual information. Moreover, no decision can be made without the proper percep-

3. Dead-reckoning here means committing to future actions before they are to be executed. The shortest human reaction time is approximately 100 msec (Boff et al. 1986). This is so-called *simple reaction time;* a boolean event where a person only has to choose between action or in-action depending on an external, pre-determined event — for example pressing a button when a light comes on. This is an appropriate measure to use here since it represents the lower limits of what can be achieved via the voluntarily controlled perceive-act cycle, and would be expected for a highly practiced skill like turn-taking. To be considered dead-reckoning, the interval from commitment to execution would then be longer than 100 msec.

tual data to base it on. There is a type of prediction besides dead-reckoning which may exist in turn-taking.[4] This kind can best be thought of as expectation or *anticipation*. We propose this as the first step towards incorporating prediction into turn-taking. This kind of prediction only affects one of the four elements of mental processing (i.e. perception — the others in our classification are cognition, decision and action), and can therefore be considered the weakest form of prediction. We hypothesize that

{H1} *Opportunities for turn-transitions are identified using a mechanism of anticipatory perceptual processing.*

Given a perceived dialogue progression P, participant A will be anticipating turn-transition T. T has associated with it a set of perceivable behavioral features F (some of which may be the traditionally called "turn signals"). Participant A will focus attention towards the occurrence of F. This he does by priming his perceptual system, thus engaging in *anticipatory perceptual processing*. We will discuss the implications of this in section 4.2, "Presentation and Interpretation: Role-Based Processing".

3.2. *Temporal Constraints in Face-to-Face Dialogue*

Face-to-face interaction is unique because it contains processes that span as much as 5 orders of magnitude of execution time, from about 100 ms (gaze, blinks), to minutes and hours (Thórisson 1999). Another way to say this is that co-temporal, co-spatial discourse contains rapid responses and more reflective ones interwoven in a complex pattern dictated by social convention. The structure of dialogue requires that participants agree on a common speed of exchange (Goodwin 1981). If the rhythm of an interaction is violated, it is expected that the violating participant make this clear to others, at the right moment, so that they can adjust to the change. For example, if a story teller suddenly forgets what comes next in her story and has to pause, she is sure to indicate this to her audience by saying something like "ahhh" or even the more explicit "Hmm, I can't seem to remember what happened next". This common speed sets an upper limit to the amount of time participants can allocate to thinking about the dialogue's form, content, and to forming responses. Newell's (1990) classification of time scales in human mental processing proposes a "cognitive band" which spans three orders of magnitude of time, from 100 ms to tens of seconds. A lot of mental processing during dialogue happens in this band, yet very few have looked at the real-time performance aspects of mental processing.

The issue of real-time is not only about speed but about *proper mental load-balancing:* ensuring that the most important processes are always run. If the story teller fails to explain the pause in her story telling, unwanted interruptions are bound to happen; the processes supporting the *delivery* of the explanation represent a higher priority than the production of the story itself. Based on Dodhiawala's (1989) principles of real-time performance, we can identify the following four aspects of real-time performance:

4. It is not clear whether by 'prediction' Sacks et al. (1974) mean ballistic action, prediction of a turn transition point in the future, or to some kind of anticipation.

1. Responsiveness: The system's (in this case dialog participant's) ability to stay alert to, and respond to, incoming information.
2. Timeliness: The system's ability to manage and meet deadlines.
3. Graceful adaptation: The system's ability to (re)set task priorities in light of changes in resources or workload, and to rearrange tasks and replan when problems arise, e.g. in light of missed deadlines.
4. Speed.

The first three are about load-balancing; the fourth requires a reference — speed compared to what? That 'what' is the real world. Following the Model Human Processor model of cognitive processing (Card et al. 1983), we can look at speed of cognition at three stages: (1) Speed of *perceptual analysis*, (2) speed of *decision*, and (3) speed of *action composition*.[5] What really matters, of course, isn't the speed of any one of these stages but that their combined output, e.g. bending down to avoid being hit in the head, are composed and executed fast enough to get the head out of the way. Of equal importance, the system has to know that at that point in time this is the most important thing to compute, both in perception and action. To achieve this feat the system has to be capable of simultaneous production of multimodal action and multimodal perception, in other words, it has to be capable of parallel processing.

To coordinate events at multiple timescales we draw on the modularization approach of behavior-based AI and hypthesize that

{H2} *The seamlessness observed in real-time turn-taking comes from the co-operation of multiple processes with different (a) update frequencies, (b) target perception-action loop times, and (c) speed-accuracy tradeoffs.*

The above real-time factors, combined with hypothesis {H2}, have lead to a modularization of YTTM that separates processes according to the urgency of their processing, the *layered feedback loop* model. Processes are load-balanced by different priority levels, and some processes can momentarily suspend processes running at other priorities. Low- and high-priority processes run in parallel, e.g. detection of interrupts, a high-frequency update process, runs in parallel with constructing narrative, a lower-frequency update process, and when the two produce results they may interfere or combine in the output behavior.

But if we have a distributed system where processes with high update rates are — by design — the first to catch a subtle, high-frequency 'interruption cue' from an interpreter, how do they communicate this fact to other processes, e.g. those in charge of telling a story? How does the goal of ignoring a presenter turn off the (modularized) ability to detect and respond to pauses? For this the processes need to have bi-directional control of each other's processing. This leads us to hypothesize that

5. Action composition is not the same as action execution (Thórisson 1997). Here we assume that execution — i.e. movement — characteristics (either simulated in graphics or actual robotics) matches roughly those of the human body.

{H3} *To support coherent output generation in a modular, distributed system, processes with different perception-action loop times are sensitive to a particular* subset *of the total set of contextual cues available, which includes perceptual data produced from input to the sensors, as well as processing states and partial output of other system elements.*

To take some (simplified) examples, the process of classifying a presenter's silence as a "hesitation", rather than a "turn signal", can rely on the context provided by other parts of the interpreter's mental processing, namely those that classify the presenter's bodily stance as *pensive*, sentence completion and semantic content as *incomplete*, and her gaze as *distracted*, all of which are bottom-up processes which support a "hesitation" theory. Other contextual cues that may influence the interpreter's behavior, given a choice between classifying behavior into a "hesitation" or a "turn signal", could be his complete lack of understanding, or perhaps his lack of something to say, which may simply result in him not taking the turn.

Now, let's look further at the layered feedback loop model.

4. PRINCIPLES OF THE YTTM

4.1. Multimodal Dialogue as Layered Feedback Loops

The YTTM follows a layered feedback loop model. The layers in this model are both *descriptive* — they are based on time-scales of actions found in face-to-face dialogue — and *prescriptive* — they specify the prioritization, or load-balancing, of computation. At each level in this model various sensory and action processes are running, primarily providing services to the level below and/or above. The highest priority is concerned with behaviors that have perceive-act cycles shorter than 1 second, typically less than 500 msec. Highly reactive actions, like looking away when you believe it's your turn to speak (Goodwin 1981) or gazing at objects mentioned to you by the presenter (Kahneman 1973), belong in this Reactive Layer. The Process Control Layer includes mental activity that relates to what we would typically categorize as the willful control of the interaction itself: starts and stops, interrupts, recognizing breakdowns, in short, everything that has to do with the *process of the dialogue* (sometimes called 'task level'). The perceive-act cycle of such events typically lie between a half and 2 seconds. Together these two layers contain the mechanisms of dialogue management, or psychosocial dialogue skills.

The lowest-priority layer, the Content Layer, is where the content or "topic" of the conversation is processed, e.g. navigating a rocket ship or cutting grass. Following hypothesis {3} (and {4} — see below), we can treat the topic knowledge residing in this layer as a black box: Its input is provided by perceptual processing in the whole system; its output is speech and multimodal behavior related to the topic of the dialogue, and actions relating to the manipulation of the topic domain.

The set of perception and decision processes actively at work in each of the three levels at any point in time is determined by several factors, one being the role of the participant at that point in time (content presenter or content interpreter). Another factor is the perception-action loop time required for the system to behave correctly. A third factor is incremental processing; multimodal, real-time interpretation is not

Table 1. The table shows which mental processes belong in each of the three priority layers of the YTTM, for content presenter and content interpreter. All tasks run in parallel, but those in the Process Control and Reactive layers have higher priority than those in the Content Layer, both in terms of processing and of execution of actions resulting from the processing. ("Process" in "Process Control" refers to the process of the dialogue, i.e. the interaction.*) This table links Figure 1, which shows the main states f the turn-taking mechansim, and Figure 2, which shows feedback loops and target loop times for each layer.*

DIALOGUE ROLE / LAYER	CONTENT PRESENTER ("speaker") PERCEPTION *(p)* & MOTOR *(m)* PROCESSES	CONTENT INTERPRETER ("listener") PERCEPTION *(p)* & MOTOR *(m)* PROCESSES
CONTENT LAYER (low priority)	Analyze interpreter's content reception *(p)* Present content *(m)*	Interpret content *(p)* Convey status of content interpretation (content-related back channel feedback) *(m)*
PROCESS CONTROL LAYER (medium priority)	Analyze interpreter's process control *(p)* Control process *(m)*	Interpret dialogue structure *(p)* Convey status of dialog structure interpretation *(m)* Control process *(m)*
REACTIVE LAYER (high priority)	Broad-stroke functional analysis *(p)* Reactive behaviors *(m)*	Broad-stroke functional analysis *(p)* Process-related back-channel feedback *(m)*

done "batch-style": There are no points in a face-to-face interaction where a full multimodal act or a whole sentence is output by one participant before being received by another and interpreted as a whole. Interpretation of multimodal input happens in parallel with multimodal output generation, continuously produced by processes running in parallel at each level.

4.2. *Presentation and Interpretation: Role-Based Processing*

Following our discussion about prediction leading up to hypothesis {H1}, we define two different sets of processes, both of which include perceptual, decision and motor tasks, that participants in a dialogue switch between depending on whether they are in the content interpreter or content presenter role.[6] We call this *role-based processing*, and it is really a kind of context sensitivity. Thus, for the period that person *A* takes the role of interpreter, one can expect him to be engaged in a set of mental activities that are different from those he is engaged in when in the role of presenter. To take an example, Goodwin & Goodwin (1986) discuss the activity of searching for a word and how this can be a cooperative activity. A content presenter may indi-

6. This does not mean that there are no processes that run during both states. Indeed, a large portion of typical perceptual, decision and motor tasks, such as glancing at the other party, smiling, etc., may run in both states.

cate to her interpreter, using gaze and body language, that she is looking for a word. The interpreter will offer to assist in the search by interjecting plausible words. Although the process is cooperative, it is the presenter who has the turn, and thereby the power to accept or reject the interpreter's suggestions (even in cases where the interpreter knows exactly what the presenter wants to convey). It is not only the relevant behavioral repertoire (visible actions) that is different for each role, but also the demands on the two participant's perceptual and decision-making systems. The roles can be thought of almost as roles in an improvisational play; they are part of the same plot but the rules for each actor's character are very different. The complication is of course that every now and then the actors switch roles according to very complex rules — they take turns.

The roles of content presenter and content interpreter are subjective: For turn-taking to work properly the concept that one participant has the turn has to be represented in the minds of all participants as a mutual belief that this participant has the turn. (They also have to share the goal of achieving efficient and cooperative communication (Grice 1989).) Moreover, their understanding of what represents appropriate moments for taking and giving turns has to also be mutually shared. Cultural differences are the clearest demonstration of how this must be so.

The concept of role-based processing can be taken one step further. According to hypothesis {H1} the process of turn-taking relies on anticipatory perceptual processing; this principle can be extended to perception *during* particular turn states. Hence, back-channel feedback, clarifications during turns, complementary gestures, additional facial expressions etc., can all be generated if needed, based on anticipatory perception. Thus, for a presenter Beth and interpreter Alan, Beth monitors Alans's behaviors (via anticipatory perception) for cues that reveal his understanding of what she is saying; Alan monitors the content of what Beth is presenting, and, via anticipatory perception, identifies places where back-channel feedback, interruptions and the like are appropriate (Table 1).

4.3. *Separating Interaction Control from Content Generation & Delivery*

One of the questions we need to answer is how the topic of the dialogue relates to the processes that control the timing and style of interaction. The representation of a topic domain is in and of itself a complex matter, and no computer model has succeeded in replicating the detailed knowledge a human has for a given domain of expertise (cf. Lenat 1995). It is unlikely that the rules for how a topic is talked about are replicated for each domain separately; it is more likely that general knowledge about how to convey information is stored once, to be reused for any topic that may be discussed. It can even be argued that this knowledge is a topic in and of itself. YTTM theory splits topic knowledge from dialogue knowledge into separate systems that talk to each other via a well-defined, small set of messages; when you are asked where you were yesterday the knowledge you use to hesitate, look up, roll your eyes, and say "hmm" is controlled by a general mechanism, separate from the processes required to fetch the piece of information required to answer the question proper. So *interaction (process) control* is separate from *content interpretation*. Turn-taking control takes into account the processing status of the domain knowledge in the same way it takes into account any other context. In this case, if it takes longer than typical for the topic knowledge system to process the input the turn-taking mechanism may decide to comment on this fact in one way or another (e.g. looking up, saying "Now, let me think..." or by direct semantic information like "Wow, that's a tough question..." — in the latter case the topic knowledge processes would communicate meta-information, i.e. that the question is 'tough', to the turn system.)

Following up on hypothesis {H3}, which claims that various modules in a distributed system are sensitive to various subsets of contextual cues, we present here the *topic-independence* hypothesis of turn-taking:

{H4} *Processing related to turn-taking can be separated from processing of content (i.e. topic) via a finite set of interaction primitives.*

This hypothesis has been informally incorporated by others (see e.g. Cahn & Brennan, 1999). In the YTTM prototype the primitives are implemented as a set of messages, exchanged via blackboards, as detailed below (see Figure 3).

4.4. *Modeling Multimodal Perception*

As a presenter, one's perceptual system is preoccupied with monitoring the progress of one's production of narrative output. But following the proposal of role-based processing, of even higher priority is distinguishing between acts of the interpreter that are insignificant to the dialogue (such as the listener casually adjusting his hair), and those that constitute communicative actions, such as a wish to interrupt. The latter behaviors take priority because they may directly affect turn-taking, and thus the course of the narration. The interpreter's top perceptual priority revolves around interpreting what the presenter is saying and making sure the presenter knows that he is following her story, giving indications of the status of his understanding processes, and interrupting when problems arise.

This emphasis on the presenter-interpreter distinction has the important result of placing the tracking of dialogue state in the driver seat among the sensory activities. It is a process that happens at the decisecond level of granularity and is highly temporally constrained. This is summarized in the following hypothesis:

{H5} *Perceptual and decision processes dedicated to tracking dialogue state have the highest priority of all mental activities related to communication.*

In other words, the processes with the highest priority in our system are perceptual processes that produce the data necessary to estimate dialogue state reliably, and the decisions related to these, which change the (mental representation of) dialogue state from one to the next. Why must this be so? Attention is a limited resource and the system has to continuously make trade-offs in processing: The faster it should be responding to a turn-taking cue, the more reliably the cue has to be detected for the interaction quality not to be degraded (increased speed means fewer "sample points" to base the decision on). Most turn-taking cues are multifaceted, involving some combination of many features such as intonation combined with a pause combined with a particular state of content production combined with a particular eye movement. The faster a decision is made in response to any perceptual cue the lower the probability that it actually represents a turn-taking cue, because the total set of events that have some *characteristics* of turn-taking cues far outnumbers that of *actual* cues, and, depending on how fast the needed cues become available, rash decisions will thus often lead to wrong decisions.

4.4.1. Functional Analysis: Characterizing the Broad Strokes 'First'

Any system that works under time-constraints and uncertainty is forced to always look at the most important data first, since time-pressure may prevent scrutiny of detail. So what constitutes the most general information for a multimodal turn-taking system? How and where do we look for it? We claim that the most significant information in conversation is the *function of discoursal actions*, and people look for it using a system of specialized processes that have a relatively high speed/accuracy ratio. These processes look at the broad strokes of the other participant's behavior 'first' — the word is in quotes because it does not imply sequential processing (all processing is parallel); the principle refers to the *priority* that functional analysis has in multimodal perception. To illustrate further, let's return to the story from the introduction.

> Alan is telling Beth an exciting story about his vacation in Nice. He presents his story through gesture and speech. Then Beth's arm starts moving and her neck stiffens. Beth's gesture is not yet recognizable as a communicative action. The movement grabs Alan's attention and keys his perceptual system in to classify the motion further, because in the next half second Alan has to make some decisions about Beth's arm movement that may affect his own behavior. Let's follow Alan's perceptual anticipation for the next 460 milliseconds . . .
>
> ***t-minus-460 msec***
>
> Beth's arm moves. Alan has to decide whether:
>
> 1: Beth's arm movement constitutes a communicative gesture, and if so,
>
> 2: what kind of gesture.
>
> Because Alan is presenting, and thus has the turn, he's reluctant to let himself be interrupted.
>
> ***t-minus-350 msec***
>
> 3: Based on Beth's expression so far, he's persuaded to pause his presentation at t-minus 350 ms (human choice reaction time is ~100 ms (Boff et al. 1986]).
>
> ***t-minus-250 msec***
>
> 4: Using Beth's gaze and the state of the dialogue, Alan decides that he will try to figure out what Beth's multimodal actions mean (i.e. what kinds of phenomena in Beth's mind does her current behavior correlate with — or serve as index of), and thus delay his presentation further.
>
> 5: Alan figures out that Beth has started making a deictic gesture (he's not sure, but "it's worth a glance") so, based on the direction of Beth's gaze, at
>
> ***t-minus-150 msec***
>
> 6: Alan looks over in the direction in which Beth is roughly pointing (where he'll see an elephant).
>
> 7: Beth's gesture becomes fully-fledged, easily recognizable deictic gesture.
>
> 8: Alan should have delayed looking. He had just reached out for his glass of beer, and now, at ***t-minus-0 milliseconds***, he sees the elephant Beth is pointing at... and knocks his beer over.

In order to conduct efficient turn-taking, Alan decided, based on the potential communicative function of Beth's actions, to pause his production of content and succumb to the turn-taking rule which states that generally a wish to interrupt should be acknowledged. Notice that had Alan continued to speak, it would either have been because he chose to do so, or that he had failed to see what Beth was doing. In other words, Alan's acknowledgment of Beth's interruption (by stopping to speak) was not delayed because he was speaking; the only way it could have been delayed was by

Alan wilfully pausing. The example illustrates that the highest-priority interpretation of a dialogue participant's behavior should not — in fact *could* not — primarily be concerned with content, for example which lexical elements can be best mapped onto a presenter's utterance, or whether an utterance at any point in time is grammatically correct, it has to be concerned with distinctions that determine broad functional strokes of behavior, i.e. extracting the features that make the major distinctions of the dialogue, *communicative* versus *non-communicative*. Computing the function of a person's behavior to mean that the person is addressing you is a necessary precursor for you to start listening; interpreting a movement's function to be a deictic one will have to happen before you can look in the direction of the pointing arm/hand/finger (or gaze) to find the referent of the action. Depending on the state of the dialogue this could either be a wish to interrupt, as in the example above, or part of conveying content. Thus, a gesture might reveal the meaning of a seemingly meaningless utterance; a nod might indicate the direction the interpreter should look for grasping the meaning of the presentation; intonation might indicate sarcasm, etc. These examples constitute broad strokes — high-level function — of behavior. The ***broad-strokes-first hypothesis*** postulates that

{H6} ***In real-time communication, analysis and interpretation of broad-stroke communicative function takes higher priority than content analysis and interpretation.***

Analysis of broad-stroke function is not the same as top-down analysis; our mental processes can use evidence from bottom-up *and* top-down to find broad stroke functions. Broad-stroke functions have to be higher priority because they provide the context for the communication itself, and by extension the presentation. On the feedback generation side, a listener's behavior of looking in the pointed direction is a sign to the presenter that he knows that her gesture is a deictic one, and that he has correctly extracted the relevant direction from the way her arm/hand/finger are spatially arranged. The gaze behavior resulting from correct functional analysis serves double duty as direct feedback, and constitutes therefore efficient process control. Thus, analysis of the contextual function of a presenter's actions and control of the process of dialogue are intimately linked through functional analysis. Furthermore, the information necessary for correct and efficient content analysis is often the necessary information for providing correct and efficient multimodal feedback behavior (Table 1).

4.4.2. Combining Multimodal Perceptual Information

Given the multiple sources of information in multimodal conversation, we are led to the following line of reasoning: A large set of aggregated cues from multiple sources of information must be more reliable than a smaller set of cues from a single mode. Thus, to achieve the most efficient trade-off between speed and accuracy of turn taking, perception related to turn-taking can be expected to draw on cues from any number of modes and sources, as long as they are informative. Therefore:

{H7} ***Reactive behaviors are based on data produced by highly opportunistic processing.***

How would such an opportunistic perceptual system combine 'evidence' from multiple sources and modes? The hypothesis we build on here is:

> **{H8}** *Separate features and cues extracted, by perceptual processes of a dialogue participant A, from a particular multimodal action by dialogue partner B, are logically combined (in the mathematical sense of the word) by other perception processes in the mind of participant A, to support generation of appropriate behavior during the interaction.*

Thus, our first approximation to this question is that the process of multimodal integration is based on boolean logic gates (cf. Duncan 1972). This has certain advantages, namely, it is easier to compute and to track the interaction of multiple boolean variables than interaction among equally many scalars, making this probably the simplest possible choice for how to combine data from multiple modes.

To relate this back to the issue of the speed/accuracy trade-off in perception and action, according to these hypotheses, the more features and modes available to someone who is assessing the dialogue (in a single perceptual analysis of turn-transitions and turn-state) the *higher the accuracy* of that perceptual assessment. In other words, estimations on part of the dialogue participants that the dialogue is in a given state, or should change to a new state, will be more accurate with an increased number of modes. This may in fact be one of the reasons why we often prefer to meet face-to-face, rather than simply talking over the phone (or sending e-mail). Paradoxically, the speed of the analysis will not be affected by the presence of more data because it is already a massively parallel process. However, increased perceptual reliability may affect the speed at which the perceiver will *act on* the extracted features. Thus, upon interpreting the multimodal act "He went [deictic manual gesture & gaze] that way," an interpreter may look sooner in the relevant direction if the manual pointing gesture is present, than if the only indication of direction is the presenter's gaze, since a manual deictic gesture is a more reliable indicator of direction than gaze alone. More efficient, speedier turn-taking can thus happen in a face-to-face meeting than any other kind. We give a working example of this in Section 5.3.1.

4.5. Decisions

"Decision" is the event where we turn a perception into a potential action — it's the switch, so to speak, for moving the body of the conversant. If you make a decision half-way (or 3/4th way or 12/27th way) you are not making a decision, you are in fact the very definition of someone who *can't* make a decision. A decision is either made or not made, a crisp event. Turn-taking decisions in YTTM are mainly made about *turn-transitions*, *provision of back-channel feedback*, and the *timing of the production of content* (i.e. when we start to say what we want to say about the content of the dialogue). Decisions in YTTM aligns with hypothesis {H8}:

> **{H9}** *A decision is based on the boolean combination of perceptual features.*

To meet our real-time demands, any *motor events* generated as a result of a decision made in the Reactive Layer has the highest priority for execution; the Process Control Layer has second priority, and the Content Layer the lowest priority. Now, before turning to the system's implementation, let's look briefly at movement generation.

4.6. Production of Motor Events

A significant part of perceptual data related to the turn-taking process lead to decisions that result in actions. A decision of a listener to interrupt the presenter may result in a series of motor events that, in the dialogue participants' culture, is a well-recognized cue for wanting to communicate something. A decision is always discrete, but in the Ymir Turn-Taking Model a decision to move e.g. a body part may or may not result in the movement actually happening: For all decisions the last stop before they become movement is controlled by a relatively monolithic action scheduler (Thórisson 1996, 1997). The action scheduler allows a decision to be cancelled up until 100 ms before its execution by committing to it only at execution time (rather than at decision time), giving YTTM the same choice reaction time found in human behavior (Boff et al. 1986).

The process of turning a (relatively) high-level decision like "interrupt the content presenter" into an acceptable series of motor events is a complex one. Canned responses are a simplification that will not work if our goal is to create a complete generative model of turn-taking; for that the model will have to be able to produce overlapping and interwoven multimodal behaviors. The YTTM uses a multimodal Motor Lexicon for turning a decision (which is a simple kind of a goal) into a motor sequence. The Motor Lexicon is a tree where the nodes are decision names such as "interruptContentPresenter" and the leaves are particular motor sequences that can signify an interrupt in the interpreter's culture. Between a node and the leaves we may have multiple branching; each branching is a named decision/goal node. At each node the motor system can choose between alternative options to achieve that decision/goal. This is where the power of the motor system comes from: Each choice can be compared to the current state of the system and agent's body, and chosen based on "goodness of fit" for that particular moment in the dialogue. For example, the decision/goal "interruptContentPresenter" may branch into the three options:

1. [raise-arm, raise-index-finger, look-at-presenter]
2. [raise-eyebrows, raise-arm, open-mouth]
3. [say-ahhh, look-at-presenter]

Each of the constituents in these three options may in turn have one or more options. If both hands of the interpreter are busy, the last option of producing speech and gaze for interrupting the presenter will be chosen by the system, since both of the other options require the arms to move. Turning the first element of the last option ("say-ahh") into motor events would result in a structure of the form:

[First: Open-mouth[*X*, *d*], Second: Produce-sound [*p*, *v*, *i*, *d*]]

Variable *X* contains a number signifying how much to open the mouth, variable *d* tells the system how fast to move, variable *p* contains the phoneme sequence "ahhh", variable *v* the volume of the utterance, and *i* the intonation pattern to use. Further details on this action scheduling scheme can be found in (Thórisson 1997).

Table 2. Prototype YTTM Covert State Deciders change turn-taking state in a real-time human-humanoid interaction. (All are ACTIVE-DURING-STATE: Dialog-On.) Rules are listed in a LISP-like syntax. For items [a]-[g] and a discussion of the rules, see Section 5.3.1.

```
STATE: I-Have-Turn                                                    1
TRANSITION TO STATE I-Give-Turn IFF:
(OR  (AND
        (I-have-something-to-say = F)
        (I-have-something-to-do = F))
  [a](AND  (Im-executing-topic-realworld-task = F)[b]
           (Im-executing-communicative-act = F)[b]
           (Other-wants-turn = T )))
```

```
STATE: I-Give-Turn                                                    2
TRANSITION TO STATE Other-Has-Turn IFF:
(OR      (Other-accepts-turn = T)[c]
         (Other-wants-turn = T)
   (AND  (Other-is-paying-general-attention = T)
         (Im-executing-topic-realworld-task = F))[b]
         (Im-executing-communicative-act = F)))[b]
```

```
STATE: I-Give-Turn                                                    3
TRANSITION TO STATE I-Have-Turn IFF:
         (Other-accepts-turn = F)
```

```
STATE: Other-Has-Turn                                                 4
TRANSITION TO STATE I-Take-Turn IFF:
(AND     (Time-since 'Other-is-presenting > 50 msec)[d]
         (Other-produced-complete-utterance = T)
         (Other-is-giving-turn = T)
         (Other-is-taking-turn = F))[e]
```

```
STATE: Other-Has-Turn                                                 5
TRANSITION TO STATE I-Take-Turn IFF:
(AND     (Time-since 'Other-is-presenting > 70 msec)[f]
         (Other-is-giving-turn = T)
         (Other-is-taking-turn = F)
      [g](OR  (Others-intonation-going-up = T)
              (Others-intonation-going-down =T)))
```

```
STATE: Other-Has-Turn                                                 6
TRANSITION TO STATE I-Take-Turn IFF:
(AND     (Time-since 'Other-is-presenting > 120 msec)[h]
         (Other-is-giving-turn = T)
         (Other-is-taking-turn = F))
```

```
STATE: I-Take-Turn                                                    7
TRANSITION TO STATE I-Have-Turn IFF:
(AND     (Other-is-paying-general-attention = T)
         (Other-is-presenting = F)
         (Other-wants-turn = F))
```

```
STATE: I-Take-Turn                                                    8
TRANSITION TO STATE Other-Has-Turn IFF:
(AND     (Time-since 'I-Take-Turn > 120msec)[i]
         (OR
          (Other-is-speaking = T)
          (Other-wants-turn = T)
          (Other-is-presenting = T)))
```

Table 3. Top-level dialogue State Deciders used in the YTTM prototype for Gandalf. States that determine whether the agent is engaged in dialogue, and therefore subsume the turn states. The dialogue states are used to set the stage for the turn-taking; dialogue starts (in this implementation) when a human being addresses the agent by name or by greeting.

STATE: ***Dialog-on*** TRANSITION TO STATE ***Dialog-off*** IFF: (AND (Other-is-saying-goodbye =T)[a] (Dialog-off = F))	**9**
STATE: ***Dialog-off*** TRANSITION TO STATES (AND ***Dialog-on I-Take-Turn***) IFF: (AND (Dialog-off = F) (OR (Other-saying-my-name =T) (AND (Other-is-greeting = T) (Other-is-addressing-me = T)) (Other-is-paying-general-attention = T)))	**10**

5. IMPLEMENTATION OF YTTM

The YTTM has been implemented in two working systems, *Puff the Magic LEGO Dragon*, and *Gandalf, the Interactive Guide to the Solar System*. The former was developed as part of a virtual world research project at LEGO, demonstrating the easy integration of YTTM with traditional planning systems, giving Puff high-level goals such as *stay-alive* and *act-playfully*, and the ability to plan actions on larger time scales than the turn (Bryson & Thórisson in press). The focus here will stay on the Gandalf prototype, with its high-fidelity multimodal perception system.

5.1. Elements of YTTM

The YTTM has been implemented by combining features from three A.I. approaches and cognitive modeling: Behavior-based A.I. (cf. Maes 1990b), the (classical A.I.) idea of blackboards and distributed processing (Adler 1989, Selfridge 1959), and the Model Human Processor (Card et al. 1983). Perception is done via a collection of *Perceptor* modules which take in sensory data, or partially processed data from other Perceptors, and compute further results. They are divided into *Unimodal Perceptors* (UPs) and *Multimodal Integrators* (MIs). UPs turn raw sensory data into more meaningful information by segmenting and processing subsets of data from a single mode (e.g. hearing or vision). This data is at or slightly above digital signal processing. MIs take processed output from UPs and other MIs and thus combine information from multiple modes when computing perception. *Decider* modules are divided into *Covert State Deciders*, which keep track of dialogue state and turns, and *Overt Deciders*, which make decisions about an agent's visible behavior. All Deciders read output from the UPs and MIs via shared blackboards. When Overt Deciders fire, they generate a Behavior Request for a particular visible behavior to happen, which are handled by an action scheduler, as explained in Section 4.6.

The Perceptors and Deciders form together the foundation of the YTTM turn-taking system. The rules and processes for the Perceptors and Deciders in the Gandalf prototype agent described here are a distillation of psychological research on the behavior found in human face-to-face dialogue (cf. McNeill 1992, Rimé & Schiaratura 1991, Pierrehumbert & Hirschberg 1990, Goodwin 1986, Kleinke 1986, Nespolous & Lecours 1986, Goodwin 1981, Kahneman 1973, Duncan 1972, Yngve 1970, Ekman & Friesen 1969, Effron 1941), and constitute a blueprint for how a turn-taking system for a particular culture (Western) can be implemented in this architecture.

5.2. *Turning Hypotheses Into Code*

Before we give a short description of the modules themselves, we will first look at how the nine preceding hypotheses relate to this YTTM implementation.

- {H1} Opportunities for turn-transitions are identified using a mechanism of anticipatory perceptual processing.

This first hypothesis is implemented by tying each Multimodal Integrator (Table 4) to a particular mental (usually turn) state, indicated in the ACTIVE-DURING-STATE slot. Turn- and dialogue states are maintained by a system of Deciders (Table 2) and their supporting Perceptors. To change a state, a Covert State Decider posts a message to a blackboard about the new state.

Multimodal Integrators (Table 4) integrate and use data from Unimodal Perceptors (Table 5). The UPs are in the Reactive Layer and take priority over all other processes. They are not linked to particular states and process continuously when data is available. A higher-level set of modules (Table 3) monitor whether the agent is actually engaged in a dialogue or not, subsuming all turn states. (None of the Deciders in

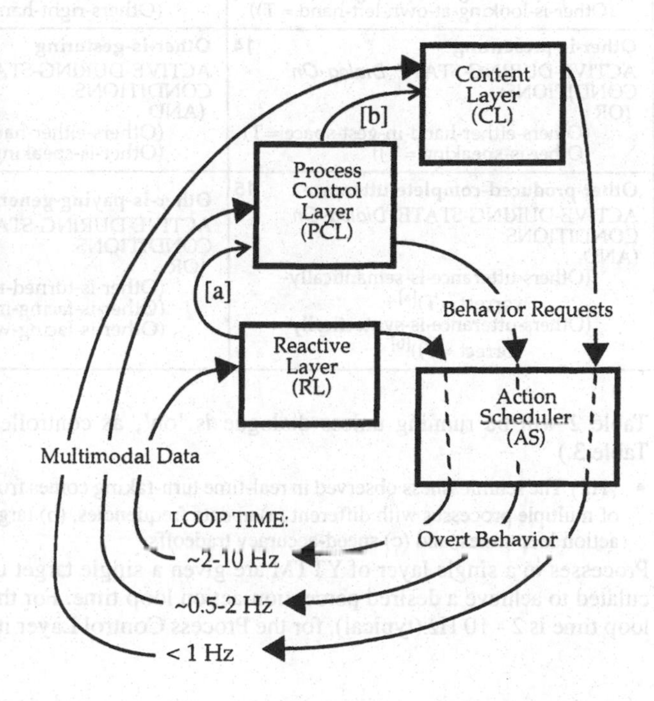

Figure 2. Multimodal input flows into all three priority layers (from Thorisson [1998]). Decision modules operate on these results and decide when to send Action Requests to an Action Scheduler, which then produces visible behavior. Target loop times for each layer is shown in Hz (compare to Table 1). It is important to note here that the frequency refers not to the layers' internal update time or sampling rate, nor to the speed of decision making, but to a full perception-action loop. [a] and [b] are partially processed multimodal data.

Table 4. Multimodal Integrators used in the Gandalf prototype using YTTM. For items [a]-[c] see Section 5.3.2.

Other-is-giving-turn **11** ACTIVE-DURING-STATE: *Other-Has-Turn* CONDITIONS: (AND (Other-is-speaking = F) (OR (AND (Other-is-looking-at-me = T) (Other-is-facing-me = T)) (AND (Other-is-looking-at-me = T) (Other-is-gesturing = F)) (AND (Other-is-gesturing = F) (Other-is-facing-me = T))))	**Other-accepts-turn** **16** ACTIVE-DURING-STATE: *I-Give-Turn* CONDITIONS: (AND (Other-is-looking-at-me = F)[c] (Other-is-presenting = T)) **Other-is-addressing-me** **17** CONDITIONS: (AND (Other-is-turned-to-me = T) (Other-is-facing-me = T) (Other-is-looking-at-me = T))
Other-wants-turn **12** ACTIVE-DURING-STATE: *I-Have-Turn* CONDITIONS: (OR (Other-is-speaking = T) (Others-hand-in-gesture-space = T))	**Other-wants-my-back-channel-feedback** **18** ACTIVE-DURING-STATE: *Other-Has-Turn* CONDITIONS: (AND (Other-is-looking-at-me = T) (Other-is-speaking = T))
Other-is-looking-at-own-hand **13** PERCEPTOR-TYPE: Multimodal-Integrator ACTIVE-DURING-STATE: *Dialog-On* CONDITIONS: (OR (Other-is-looking-at-own-right-hand = T) (Other-is-looking-at-own-left-hand = T))	**Others-either-hand-in-gest-space** **19** PERCEPTOR-TYPE: Multimodal-Integrator ACTIVE-DURING-STATE: *Dialog-On* CONDITIONS: (OR (Others-left-hand-in-gesture-space = T) (Others-right-hand-in-gesture-space = T))
Other-is-presenting **14** ACTIVE-DURING-STATE: *Dialog-On* CONDITIONS: (OR (Others-either-hand-in-gest-space = T) (Other-is-speaking = T))	**Other-is-gesturing** **20** ACTIVE-DURING-STATE: *Dialog-On* CONDITIONS: (AND (Others-either-hand-in-gest-space = T) (Other-is-speaking = T))
Other-produced-complete-utterance **15** ACTIVE-DURING-STATE: *Dialog-On* CONDITIONS: (AND (Others-utterance-is-semantically-correct = T)[a] (Others-utterance-is-syntactically-correct = T))[b]	**Other-is-paying-general-attention** **21** ACTIVE-DURING-STATE: *Dialog-Off* CONDITIONS: (OR (Other-is-turned-to-me = T) (Other-is-facing-me = T) (Other-is-facing-workspace = T))

Table 2 will be running unless dialogue is 'on', as controlled by the Deciders in Table 3.)

- {H2} The seamlessness observed in real-time turn-taking comes from the co-operation of multiple processes with different (a) update frequencies, (b) target perception-action loop times, and (c) speed-accuracy tradeoffs.

Processes in a single layer of YTTM are given a single target update frequency, calculated to achieve a desired perception-action loop time. For the Reactive Layer this loop time is 2 - 10 Hz (typical), for the Process Control Layer it is 0.5 - 2 Hz, and for

Table 5. Unimodal Perceptors used in the Gandalf prototype. Some features of the user's behavior are computed continuously, these are referred to with variables filling the INDEX and DATA slots; other data are only computed when a particular module runs (i.e. when needed), necessitating a function call (bold italic). For discussion see Section 5.3.3.

Other-is-facing-workspace 22 TYPE: Unimodal-RL-body-perceptor DATA-1: workspace DATA-2: **get-head-direction** FUNC: ***x-facing-y***	**Other-is-looking-at-me** 30 TYPE: Unimodal-RL-body-perceptor DATA-1: my-own-face DATA-2: **get-gaze-direction** FUNC: ***x-looking-at-y***
Other-is-looking-at-own-right-hand 23 TYPE: Unimodal-RL-body-perceptor DATA-1: **get-gaze-direction** DATA-2: **get-r-wrist-position** FUNC: ***u-looking-at-hand?***	**Others-right-hand-is-in-gesture-space** 31 TYPE: Unimodal-RL-body-perceptor DATA-1: **get-r-wrist-position** DATA-2: **get-trunk-direction** FUNC: ***hand-in-gest-space?***
Other-is-looking-at-own-left-hand 24 TYPE: Unimodal-RL-body-perceptor DATA-1: **get-gaze-direction** DATA-2: **get-l-wrist-position** FUNC: ***u-looking-at-hand?***	**Others-left-hand-is-in-gesture-space** 32 TYPE: Unimodal-RL-body-perceptor DATA-1: **get-l-wrist-position** DATA-2: **get-trunk-direction** FUNC: ***hand-in-gest-space?***
Others-syntax-is-complete 25 TYPE: Unimodal-RL-speech-perceptor INDEX: last-utterance DATA: others-word-stream FUNC: ***syntax-complete?***	**Other-is-speaking** 33 TYPE: Unimodal-RL-prosody-perceptor INDEX: 40-msec-chunk DATA: others-audio-stream FUNC: ***x-speaking?***
Other-is-turned-to-me 26 TYPE: Unimodal-RL-body-perceptor DATA-1: my-own-face DATA-2: **get-trunk-direction** FUNC: ***turned-to?***	**I-see-other** 34 TYPE: Unimodal-RL-vision-perceptor INDEX: body-socket-connection DATA: *socket-object1* FUNC: ***visual-connection-alive?***
Other-is-turned-to-workspace 27 TYPE: Unimodal-RL-body-perceptor DATA-1: workspace DATA-2: **get-body-direction** FUNC: ***x-facing-y***	**Other-is-facing-me** 35 TYPE: Unimodal-RL-body-perceptor DATA-1: my-own-face DATA-2: **get-head-direction** FUNC: ***x-facing-y***
Others-utterance-is-grammatically-correct 28 TYPE: Unimodal-RL-speech-content-perceptor INDEX: last-utterance DATA: others-word-stream FUNC: ***grammar-complete?***	**Others-utterance-is-semantically-complete** 36 TYPE: Unimodal-RL-speech-con tent-perceptor INDEX: last-utterance DATA: others-word-stream FUNC: ***semantics-complete?***
Others- intonation-is-going-up 29 TYPE: Unimodal-RL-prosody-perceptor INDEX: 300-msec-chunk DATA: others-audio-stream FUNC: ***inton-direction***	**Others-intonation-is-going-down** 37 TYPE: Unimodal-RL-prosody-perceptor INDEX: 300-msec-chunk DATA: others-audio-stream FUNC: ***inton-direction***

the Content Layer it is 1 Hz and slower (Figure 2). As mentioned earlier, processes in the Reactive Layer have the largest speed/accuracy ratio, those in the Content Layer have the highest accuracy and lowest relative speed, with those in the Process Control Layer in the middle. All UPs are in the RL in our implementation, while other types of modules are found in all layers.

- {H5} Perceptual and decision processes dedicated to tracking dialogue state have the highest priority of all mental activities related to communication.
- {H6} In real-time communication, analysis and interpretation of broad-stroke communicative function takes higher priority than content analysis and interpretation.

By placing all low-level processes in the Reactive Layer, highest priority is given to processing raw data that can help determine the dialogue state (Table 5). Topic interpretation happens entirely in processes situated in the Content Layer, the lowest-priority layer in the system. A scheduling system ensures a guaranteed processing time for the Reactive and Process Control layers. This varies between deployment platforms, and is tuned based on the target loop-times.

- {H3} To support coherent output generation in a modular, distributed system, processes with different perception-action loop times are sensitive to a particular subset of the total set of contextual cues available, which includes perceptual data produced from input to the sensors, as well as processing states and partial output of other system elements.

As we established in sections 2. and 4., the human mind interprets the world incrementally. For example, the movement of an object becomes available in a perceiver's mind sooner than its color (Kosslyn & Koenig 1992). This means that at any point in time some mental processes contain partial information about the world. To make use of this partial information the intermediate stages of data should be made available to other processes, in case they need it or are able to use it. This is an ideal problem to solve with a blackboard architecture. There are two blackboards used for perception in YTTM; they sit between the three layers. Bottom-up processing pushes incrementally more detailed sensory data upwards, from UPs to MIs to more knowledge-intensive processes; decisions and planning from the deliberative processes in the Content Layer push expectations and anticipatory commands downward, all via the blackboards. Any module that needs a particular set of data to produce its output looks at one of the two blackboards. If it finds the data it needs, it processes it and places the results on one of the two blackboards, making them available to other modules.

Topic-Knowledge-System-Received-Speech-Data
Speech-Data-Available-For-Analysis
Topic-Knowledge-System-Parsing-Speech-Data
Topic-Knowledge-Ssystem-Successful-Parse
Content-Layer-Action-Available
I-Have-Reply-Ready
Topic-Knowledge-System-Real-World-Action-Available
Im-Executing-Topic-Speech-Task
Im-Executing-Topic-Realworld-Task
Im-Executing-Topic-Multimodal-Act
Im-Executing-Topic-Communicative-Act
Im-Executing-Communicative-Act

Figure 3. A basic set of communication primitives from processes in the Process Control Layer to a Topic Knowledge Base in the Content Layer, posted on the blackboard shared by CL and PCL. When these are posted they are timestamped, and provided with a pointer that allows other modules to access the data that the message refers to. The primitives form part of the turn-system's contextual cues. The list must be extended for domains more complex than the one explored here.

- {H4} Processing related to turn-taking can be separated from processing of content (i.e. topic) via a finite set of interaction primitives.

Dialogue and turn states are tracked in the Process Control Layer; knowledge systems related to the topic are placed in the Content Layer and handle everything

Table 6. Overt Decision Modules used in Gandalf's Reactive Layer. *These modules control Gandalf's reactive behavior. For discussion see Section 5.3.4.*

Show-Im-taking-turn **38** EL: 5000 msec BehaviorRequest: Show-im-taking-turn FIRE-CONDS: (I-take-turn = T) RESTORE-CONDS: (I-take-turn = F)	**Show-Im-giving-turn** **43** EL: 2000 msec BehaviorRequest: Show-Im-giving-turn FIRE-CONDS: (I-give-turn = T) RESTORE-CONDS: (I-have-turn = F)
Show-I-know-other-is-addressing-me-1 **39** EL: 200 msec BehaviorRequest: Smile POS-CONDS: (Im-executing-speech-act = T) NEG-RESTR-CONDS: (Other-is-turned-to-me = F)	**Show-Im-giving-turn-2** **44** EL: 2000 msec BehaviorRequest: Show-Im-giving-turn FIRE-CONDS: (I-give-turn = T) RESTORE-CONDS: (I-give-turn = F)
Show-I-know-other-is-addressing-me-2 **40** EL: 200 msec BehaviorRequest: Eyebrow-greet POS-CONDS: (AND (Other-is-saying-my-name = T) (Other-is-turned-to-me = T) (Other-is-facing-me = T)) RESTORE-CONDS: (Other-is-turned-to-me = F)	**Show-Im-listening** **45** EL: 200 msec BehaviorRequest: Brows-in-pensive-shape FIRE-CONDS: (AND (Other-is-saying-my-name = T) (Other-is-turned-to-me = T) (Other-is-facing-me = T)) RESTORE-CONDS: (Other-is-turned-to-me = F)
Initialize-dialogue **41** EL: 200 msec BehaviorRequest: Face-neutral FIRE-CONDS: (Dialog-On = F) RESTORE-CONDS: (Dialog-On = T)	**Show-I-know-other-is-not-addressing-me** **46** EL: 1000 msec BehaviorRequest: (***Turn-to*** 'Work-space) FIRE-CONDS: (AND (Dialog-On = F) (Other-is-turned-to-me = F)) RESTORE-CONDS: (Other-is-taking-turn = T)
Look-puzzled-during-awkward-pause **42** EL: 1000 msec BehaviorRequest: Look-puzzled FIRE-CONDS: (AND (other-is-turned-to-me = T) (other-is-facing-me = T) (***Time-since*** 'Other-is-facing-me > 400)) RESTORE-CONDS: (Other-is-turned-to-me = F)	**Look-aloof** **47** EL: 1000 msec BehaviorRequest: Look-aloof FIRE-CONDS: (AND (Other-is-turned-to-me = T) (Other-is-facing-me = T) (***Time-since*** 'Other-is-facing-me > 800) (Dialog-On = T) (Other-is-speaking = F) (Topic-Knowledge-System-Parsing-Speech-Data = F) RESTORE-CONDS: (Other-is-turned-to-me = F)

related to the topic. Topic and dialogue processes talk to each other via a limited set of messages (Figure 3).

- {H7} Reactive behaviors are based on data produced by highly opportunistic processing.

To meet this objective, the processing necessary for producing coherent system behavior — including content and gesture analysis — is sliced into small units, each responsible for only a fraction of the overall interpretation. These processes are distributed throughout the three priority layers.[7] Functions for Unimodal Perceptors and Multimodal Integrators (values in slot 'FUNC' in Table 5, bold italic values in tables 2, 6 and 7) provide services that describe various parts of the dialogue state and outside world at any moment in time. Since it is not possible to predict which pieces of

---

7. Processing related to constructing a morphology (motor program) based on decisions and goals (such as 'greet-other') is done by the action scheduler (Section 4.6) running in parallel with processes in each of the layers, at the same priority as processes in the Reactive Layer.

the data will be available at any moment in time, the small units make opportunism the default method by which the system does interpretation and produces behavior.

All modules contain a list of conditions, in their FIRE-CONDS slot, whose boolean combination determines their output to the blackboards, meeting the last two hypotheses:

- {H8} Separate features and cues extracted, by perceptual processes of a dialogue participant A, from a particular multimodal action by dialogue partner B, are logically combined (in the mathematical sense of the word) by other perception processes in the mind of participant A, to support generation of appropriate behavior during the interaction.
- {H9} A decision is based on the boolean combination of perceptual features.

5.3. *Perceptor & Decider Rules*

This section explains selected modules, where their rules come from and how they interact. As mentioned before, this information is specific to a culture, and is included here as a reference for how rules are encoded in this implementation of the YTTM for supporting the interactive Gandalf character. We start with the high-level turn-rules and trace our steps backwards, ending with the Deciders that turn perceptual and cognitive data into situated behavior.

5.3.1. *Turn States (Table 2)*

The turn states in Table 2 form the crux of the turn-taking system in this implementation of the YTTM. These relatively reactive modules help determine the agent's 'interactive personality'. For example, by removing the second condition in Decider 1 ([a] in Table 2) an agent will not give the turn if it is engaged in real-world tasks, even if the other wants the turn. As seen in Transition Rule 4, the transitions are modeled explicitly as states; *I-Take-Turn* leading into *I-Have-Turn*. The conditions marked [b] are determined by the topic knowledge system and form part of the system's contextual cues. Condition [c], Other-accepts-turn, is a perception, not a state; the state *Other-Has-Turn* is thus driven off the perceptual system.

The function ***Time-since*** takes two variables, a status message (e.g. Other-is-speaking) and a time in milliseconds. In case [h] (module 6), for example, it will return *true* if the time since the user stopped presenting (according to the agent's perceptual processing) is greater than 120 msec. The upper bound on the acceptable pause between one partner giving the turn and the other taking it (again, in the Western world) is 250 msec or less (Goodwin 1981). This observation has been implemented as three transition rules 4, 5, and 6, gradually less strict in their necessary conditions. The first rule requires a complete utterance to have been produced (measured by whether the utterance is syntactically correct and if it makes sense[8]) for the agent to take turn. However, given such evidence, the probability of a valid turn transition is so high that a mere 50 msec [d] wait is sufficient. The second rule does not require a complete utterance, but uses intonation direction as an indication [g], which will be

8. The semantic completeness of a multimodal event is computed in various ways from context. For example, should a user utter the words "what is that?" with no accompanying body movements the semantic completeness is given a lower score than if that utterance had been complemented by a glance and/or a deictic gesture that singles out a relevant object.

Table 7. Overt Decision Modules used in Gandalf's Process Control Layer. *For more details see Section 5.3.4.*

Acknowledge-others-attention-during-presentation **48** EL: 20000 msec BehaviorRequest: (***Gaze-At*** 'Other) FIRE-CONDS: (AND (Im-executing-act = T) (Other-is-looking-at-me = T)) RESTORE-CONDS: (I-take-turn = T)	**Show-Im-done-with-task-by-looking-at-other** **54** EL: 20000 msec BehaviorRequest: (AND (***Turn-head-To*** 'Other) (***Gaze-At*** 'Other)) FIRE-CONDS: (Im-executing-topic-realworld-task = F) RESTORE-CONDS: (Im-executing-topic-realworld-task = T)
Turn-to-other-when-I-speak **49** EL: 20000 msec BehaviorRequest: (***Turn-head-To*** 'Other) FIRE-CONDS: (Im-executing-speech-act = T) RESTORE-CONDS: (Im-executing-communicative-act = T)	**Look-at-domain-with-other** **55** EL: 20000 msec BehaviorRequest: (***Turn-head-To*** 'workspace) FIRE-CONDS: (AND (Other-is-facing-domain = T) (I-have-turn = T) (Other-is-speaking = F)) RESTORE-CONDS: (Other-has-turn = T)
Hesitate-during-delay-in-reply-formulation **50** EL: 500 msec BehaviorRequest: Show-Hesitation FIRE-CONDS: (AND (Dialog-On = T) (I-have-turn = T) (Speech-data-available-from-other = T) (***Time-since*** 'Other-is-speaking > 70 msec) (I-have-reply-ready = F) (other-is-speaking = F)) RESTORE-CONDS: (I-give-turn = T)	**Allow-other-to-interrupt-me-during-task** **56** EL: 20000 msec BehaviorRequest: (***Turn-head-To*** 'Other) FIRE-CONDS: (AND (Im-executing-topic-realworld-act = T) (Other-is-looking-at-me = T) (other-is-facing-me = T) (Other-is-speaking = T)) RESTORE-CONDS: (Im-executing-topic-realworld-task = F)
Pay-attention-to-my-own-action **51** EL: 20000 msec BehaviorRequest: (***Turn-head-To*** 'Work-space) FIRE-CONDS: (Im-executing-topic-realworld-task = T) RESTORE-CONDS: (Im-executing-topic-realworld-act = F)	**Show-Im-idle** **57** EL: 20000 msec BehaviorRequest: Restless FIRE-CONDS: (Other-is-facing-me = F) RESTORE-CONDS: (Other-is-facing-me = T)
Show-Im-listening-to-other **52** EL: 20000 msec BehaviorRequest: (AND (***Turn-head-To*** 'Other) (***Gaze-At*** 'Other)) FIRE-CONDS: (AND (Other-is-speaking = T) (Other-is-paying-general-attention = T)) RESTORE-CONDS: (I-have-turn = T)	**Turn-to-other-when-I-present** **58** EL: 2000 msec BehaviorRequest: (***Turn-Head-To*** 'Other) FIRE-CONDS: (Im-executing-topic-communicative-act = T) RESTORE-CONDS: (Im-executing-topic-communicative-act = F)
Acknowledge-other-is-addressing-me **53** EL: 2000 msec BehaviorRequest: (***Turn-To*** 'Other) FIRE-CONDS: (AND (Im-executing-topic-realworld-task = T) (Other-is-looking-at-me = T) (Other-is-facing-me = T) (Other-is-speaking = T)) RESTORE-CONDS: (Im-executing-topic-realworld-task = F)	

going down ("final fall") if the user is stating a command and up ("final rise") if the user asked a question (Thórisson, in press, Pierrehumbert & Hirschberg 1990). In this rule the agent waits 70 msec before acting [f] (20 msec longer than in module 4). The third rule catches the condition when the user's utterance is *not* complete (or takes longer than 70 msec to be computed) and intonation is not determinate or is not computed. In this case taking the turn is delayed for an additional 70 msec, bringing the wait up to 120 msec [h]. This breakdown exemplifies how *cascaded decision modules* can be used to track state, and use real-time as part of the processing. It's also an example of our discussion in Section 4.4.2 about the reliability of perceptual data and how soon it can be acted on. Notice that the conditions in all but the last rule work as "evidence" of a certain state of the world being true. Even if the first two rules fail, the third will default to true, unless outside events cause other states in the mean time. If module 6 fails other rules will fire to stabilize the system (see examples below).

Since the conditions *Other-is-taking-turn* and *Other-is-giving-turn* are both perceptual & transitional states, and thus measured by separate, independent perceptual processes, the condition can arise that they are both true ([e] in Table 2). This might happen if for example a non-native speaker uses different rules of conduct when taking turns. Since they are not mutually exclusive, both have to be listed here for higher certainty that the perceived state is correct.

Two rules deal with collaborative mistakes: Transition 3 happens if the agent gives turn but the partner shows no signs of accepting it (time delays may be needed to prevent premature firing of this module). Transition 8 only happens if the agent mistakenly took the turn. If the time since the agent's decision to take turn is greater than 120 msec and the user is still talking or seems to be wanting the turn, the turn transition was possibly made erroneously by the agent ([i]). This might be caused by a failure in the agent's perceptual or decision mechanisms, or because the presenter reversed a decision to give turn.

5.3.2. Multimodal Integrators (Table 4)

In Multimodal Integrator 21, any of the conditions will trigger an *Other-is-paying-general-attention* message to get posted to a blackboard (all OR states are inclusive). Integrator number 13 is used to flag the potential presence of iconic gestures (manual gestures where the hand plays the role of another object (Rimé & Schiaratura 1991, Effron 1941)) and thus prime the knowledge system to analyze its meaning.

Research shows that when interpreters intend to take turn (again, in the Western world) when given by the presenter, they pull away their gaze, which typically was focused on the presenter's face up until that point (Kahneman 1973, Duncan 1972). We have captured this in its simplest form in module 16, where the rule Other-is-looking-at-me = F ([c] in Table 4) requires the other's gaze to fall elsewhere than on the agent's face.

5.3.3. Unimodal Perceptors (Table 5)

The UPs provide all medium and low-level perception necessary, and thus form the foundation for all decisions about turn-taking and covert behavior. The function of Unimodal Perceptors such as module 25 uses complex algorithms and may not always compute the answer fast enough for the dialogue to proceed correctly. Proper load-balancing via the layers, along with flexible rules, is critical to achieve the required realtime performance of the whole system: Should the UPs not get sufficient processing cycles, interaction will suffer.

5.3.4. Overt Decision Modules (Tables 6, 7)

The Overt Decider modules have the role of generating visible behavior in response to dialogue events and the agent's own mental events. When their conditions are met they fire a Behavior Request to the action scheduler. After firing they wait for the conditions in RESTORE-CONDS to become true; until then they are unable to fire.

The agent's behavioral system needs to know the time-dependency of every decision and plan made, because one decision may interfere with another, and have to be put on hold. One way this is handled is via scheduling prioritization based on the three layers, a second way is via time-dependency: Each decision made by the system has a time-out associated with it that determines how long it may be buffered in the system, waiting to be executed. In our implementation this is done via the Expected Lifetime (EL) variable, whose value determines how long a Behavior Request produced by an Overt Decider can live in the system without being turned into motor movements. If the EL time is reached before the act can be executed (for one reason or another) the Behavior Request is cancelled. EL values are selected based on psychological studies, but tuned empirically.

Research has shown that greetings are often accompanied by widening of eyes and/or a brief lifting of eyebrows (Schegloff & Sacks 1973). This has been implemented in Decider 40. Goodwin (1981), Duncan (1972) and others have shown that when taking the turn people glance away from their partner, which has led to module 38: The behavior request Show-Im-taking-turn can be realized in one of two ways by the action scheduler: (1) Raising the eyebrows and quickly glancing to the side and back, or (2) turning to face the other, quickly glancing to the side and back, and opening the mouth slightly. In Western cultures the possibilities are of course not limited to these two, and one can imagine a system where, for each such behavior, several variations exist. (Variations can continue to be added to this system, potentially into the hundreds.) The accompanying decision to look at the content presenter to show attention is encapsulated in module 52.

When executing a domain act, Gandalf will look at the events it causes, module 51. In Western turn-taking a presenter tends to look back at the interpreter when he's done presenting, signalling that the turn is available (Goodwin 1981). This is captured in module 54.

To indicate problems or delays in topic processing, module 50 will jump in and display a hesitation. In the prototype a hesitation consisted of three realizations: (1) Saying "ahhh...", (2) gazing upwards, or (3) putting on a pensive facial expression. The EL for this module is 500 msec. This means that if Gandalf decides to hesitate, but then 500 msec pass and this decision is not realized as movement (for whatever reason), the decision will not get realized at all. When this happens, instead of monitoring this event internally, we let the real-world loop catch the result: If Gandalf failed to hesitate and the user took back the turn as a result, Gandalf's turn-taking mechanism would sense this condition via perceptual mechanisms and give back the turn, and thus pick up the slack.

For spatial behaviors that are made relative to real-world objects (most movements besides facial expression) we needed special methods. For example, since the position of the other participant changes, as does the direction of the agent's head, the relative difference needs to be measured each time the agent wants to look at the other. For this we implemented a function that returns the difference between two objects and uses it to calculate where to move the agent's motors. An example of this is module 48 (the condition Im-executing-act refers to any act, whether it originated in the topic knowledge system or not). Module 52 shows the use of two functions, one for turning the head, one for changing the gaze; this module contains a reference to realworld-task which is true whenever Gandalf travels around the virtual solar system.

5.4. System Setup & Performance

5.4.1. Sensory & Display Hardware & Software

The prototype system for Gandalf consists of eight computers: Four computers are dedicated to sensation; one each for prosody analysis, gaze calculation, geometric body modeling, and speech recognition (Thórisson, in press). Two computers are used to run the Multimodal Integrators, Deciders, topic knowledge system, and motor scheduling. Two computers manage the animation for the face and for the three-dimensional model of the solar system (Figure 4). The display showing Gandalf's head and hand ("Agent screen") is angled in such a way as to allow him to look at the solar system (to his left) and see the human user right in front. The human sees the workspace display right in front of her and Gandalf to her left.

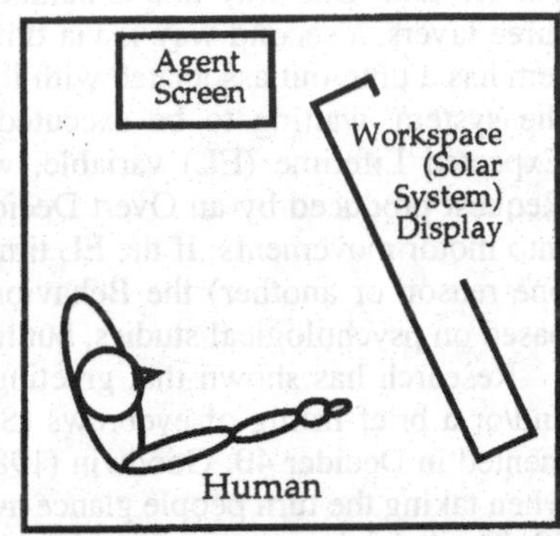

Figure 4. Top view of display layout in the Gandalf prototype setup.

5.4.2. Performance

The Ymir Turn-Taking Model as implemented shows a remarkable flexibility and adaptability considering its relatively small rule base. A large part of that flexibility we believe comes from the layered approach, as well as the explicit handling of time in the system. The number of people who have interacted with Gandalf, and it's cousin Puff, is in the hundreds; most of them have only been given general instructions such as "act as if you are interacting with another person". When Gandalf senses that the human is present, a greeting, such as "I am Gandalf, your guide to the Solar System; I can fly to the planets and tell you about them" sets up the right expectations with regard to the task and guides users to ask the questions that Gandalf understands. Gandalf's sensory and turn-taking mechanisms presented above make sure that the greeting is uttered at the right time — rarely does it fail. Within less than a minute people are communicating naturally and taking turns efficiently, flying to the planets and listening to and watching Gandalf talk about them and their moons. Interacting with Gandalf surely requires no manual. In questionnaires collected from users after talking to Gandalf, users give the system very high grades on interactivity, speech understanding, speech generation, and intelligence (this is probably not as much a reflection of the system as it is of people's perception of it). Scales are grounded in each end by asking people to compare the system to interactions with animals, such as fish, dogs, and cats on the one hand, and humans on the other. Grades on these scales for Gandalf typically fall somewhere betwen the interactivity of a dog and a real human. Interestingly, when the mechanisms presented in this chapter are turned *off*, compared to an idential version with only verbal responses to user's questions, a (statistically) significant decrease in Gandalf's scores for language understanding and language expression is observed, moving closer to the score for these parameters given for human-dog interaction. When all of the dialogue intelligence is turned on people score Gandalf's language abilities significantly closer to human-human interaction.

Figure 5 is a randomly selected five-second segment from a corpus of hours of interaction recordings, showing how the turn-taking system typically performs in interaction with real users. The user was looking and facing Gandalf during the segment. A subset of the modules in the full system are plotted. In this example a Multimodal Integrator for sensing that the other is giving turn (Other-is-giving-turn, module 11 in Table 4) failed on the second turn transition [a] (it worked correctly in case [b]). Nontheless, the system performed correctly because these two perceptual sensors are not mutually exclusive, and in this case the output from the sensor for user taking turn resulted in acceptable behavior. In all cases in this segment the agent gives and takes turn under 70 msec; they turn out to be correctly estimated transitions as well. This is an example of the architecture allowing the system to tend towards homeostasis in the presence of error, and how a behavior-based architecture results in real-time performance while preserving structural simplicity.

Figure 6 shows another example of a typical event sequence, chosen to demonstrate further the internal events that form part of the context of the dialogue, along with external states and events. The system recognizes that the user is taking the turn [a] and gives turn [b], and shows that it's giving turn [c]. The user starts speaking about 50 msec later. His request is "Tell me about that planet" (pointing at the screen). The system is relatively slow to take turn again [d] (approx. 500 msec), and shows that its taking it [e]. The duration since the user stopped speaking and gave the turn is now long (relatively speaking) and the system decides that it should show

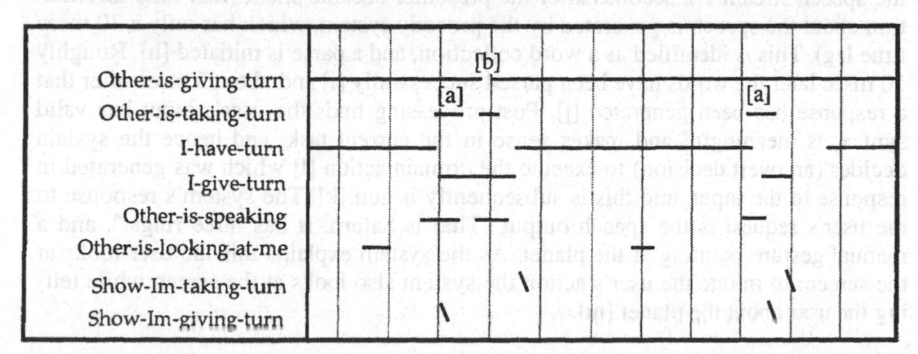

Figure 5. Example of turn-taking performance by YTTM. Each vertical line marks one second. Horizontal line means condition on left hand side was True *during that period. Decisions for showing that the agent is taking and giving turn (resulting from state changes) are plotted in the bottom two lines. For further explanation, see Section 5.4.2.*

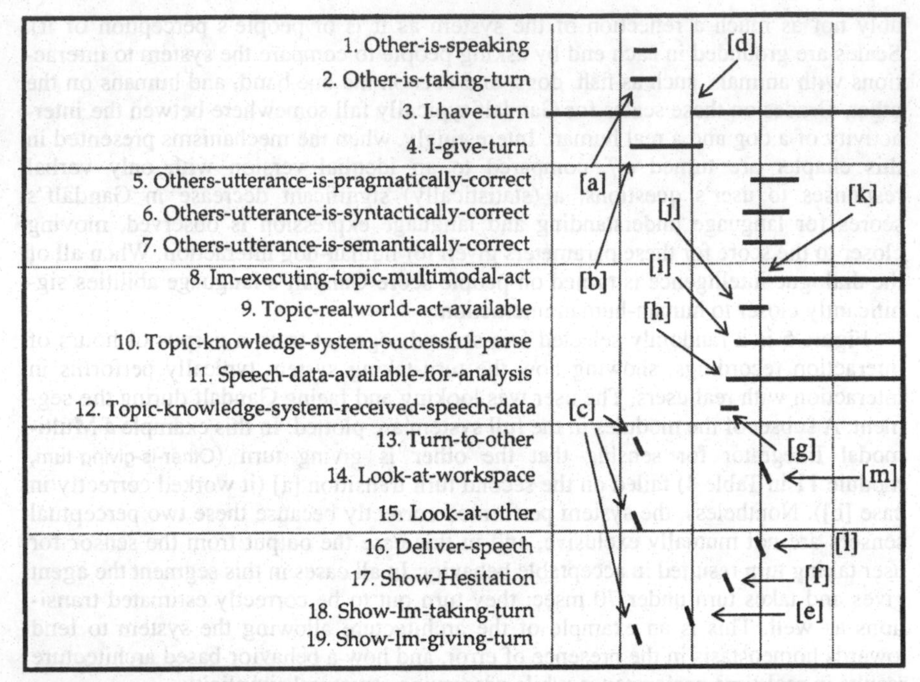

Figure 6. Example of turn-taking events showing more features of the internal contextual messages, in lines 5-11. The figure shows partial representation of internal variables during run-time. Horizontal lines mark seconds. See text for discussion.

hesitation [f], since it has accepted the turn but not yet started to respond to the content of what the user said (in fact, as of this moment, the system has no idea *what* the user said, just *that* the user said ... something). A few milliseconds later the topic knowledge system receives a report from the speech recognition subsystem [g]. In this implementation the speech recognition usually delivered words extracted from the speech stream 1-2 seconds after the presenter became silent (real-time information about the speech is generated by the prosody system, which has only a 30 msec time lag). This is identified as a word collection, and a parse is initiated [h]. Roughly 10 msec later the words have been parsed successfully [i] and about 5 msec after that a response has been generated [j]. Post-processing finds the user's input has valid syntax, is meaningful and makes sense in the current task, and hence the system decides (an overt decision) to execute the domain action [l] which was generated in response to the input, and this is subsequently begun [k]. The system's response to the user's request is the speech output "That is Saturn. It has three rings.", and a manual gesture pointing at the planet. As the system explains this the user looks at the screen; to mimic the user's action the system also looks at the screen while telling the user about the planet [m].

5.5. *Discussion & Summary*

The hypotheses presented in this chapter formalize elements needed for a complete generative model of real-time, face-to-face turn-taking. They provide a foundation for the construction of a turn-taking model, YTTM, that has been implemented using completely automated perception of a dialogue participant's behavior, including speech, prosody, gesture and body language, and generating real-time animation and speech output. In goal-directed dialogue with humans the implemented model, in the form of a humanoid agent, produces turn- and dialogue behavior very similar to that seen in human-human dialogue.

The model described combines classical AI and behavior-based AI by proposing a structure for the two to interact to achieve both real-time behavior and long-term planning. It has done so using a particular modularization based on perception-action loop times, along with a set of message types for coordinating domain knowledge with interaction knowledge. YTTM covers the complete loop from perception to action, and, as such, bridges semantic analysis, situated dialogue, discourse structure (Clark 1992), auditory perception, computer vision (Thórisson, in press), and action selection (Thórisson 1997).

The prototype described implements a number of rules based on selected psychological research spanning the last 60 years. More of these rules could be added to create a larger repertoire of perceptual states and response types in this prototype. Undoubtedly this would make the agent capable of dealing better with various boundary conditions — it currently has few error recovery mechanisms — and with conditions such as interruptions, hesitations, restarts, reformulations, and an interaction with a higher degree of mixed-initiative dialogue. We also believe that giving the turn system the ability to do strong prediction, 1-2 seconds into the future, would greatly enhance its interactive intelligence. As the examples of performance show however, the model performs reasonably well in typical situations.

Needless to say, given the tall order of creating a natural ("manual free") interactive dialogue system, significant testing remains to be done to map out the boundaries and limitations of the YTTM. This is not straightforward since any such model postulates dependencies on context: Internal knowledge states, external real-time events, pending plans and current states of the bodies of the participants, as well as their overarching goals. The hypotheses proposed are well suited for empirical testing of how the model relates to human cognitive mechanisms. Future work includes extensions to the agent's knowledge and domain action capabilities, as well as giving it a full body. To explore the practical applications of YTTM it can be employed in settings with greatly varying sensory capabilities, such as keyboard and mouse, speech-only, and speech, mouse and keyboard. These, and other variations on the model, are currently being explored with promising initial results.

5.6. *Acknowledgments*

I did the bulk of this research while at the M.I.T. Media Lab and LEGO Digital. I am grateful for their support, and the other sponsors of this work: HUMANOID sf., Thomson-CSF and TSG Magic. The work owes a lot to interactions with Richard A. Bolt, Steve Whittaker, Tom Malone, Lynn Walker, Justine Cassell, and last but not least, Pattie Maes. To them I am thankful. I would also like to thank my colleagues

Chris Johnson, Rich Cullingford, Inger Karlsson and John DiPirro for helpful comments on this paper.

6. REFERENCES

Adler, R. (1989). Blackboard Systems. In S. C. Shapiro (ed.), *The Encyclopedia of Artificial Intelligence*, 2nd ed., 110-116. New York, NY: Wiley Interscience.

Boff, K. R., L. Kaufman, & J. P. Thomas (eds.) (1986). *Handbook of Human Perception*. New York, New York: John Wiley and Sons.

Bryson, J. & K. R. Thórisson (in press). Dragons, Bats and Evil Knights: A Chatacter-Based Approach to Constructive Play. Submitted to *Virtual Reality, Special Issue on Intelligent Agents*. London: Springer.

Cahn, J. E. & S. E. Brennan (1999). A Psychological Model of Grounding and Repair in Dialog. *Proceedings of the Fall 1999 AAAI Symposium on Psychological Models of Communication in Collaborative Systems*, Sea Cliff, Massachusetts, November 5-7, 25-33.

Card, S. K., T. P. Moran, & A. Newell (1983). *The Psychology of Human-Computer Interaction*. Hillsdale, New Jersey: Lawrence Earlbaum Associates.

Cassell, J. & K. R. Thórisson (1999). The Power of a Nod and a Glance: Envelope vs. Emotional Feedback in Animated Conversational Agents. *Applied Artificial Intelligence*, 13 (4-5), 519-538.

Clark, H. H. (1992). *Arenas of Language Use*. Chicago, Illinoi: University of Chicago Press.

Clark, H.H. & E. F. Schaefer (1989). Contributing to Discourse. *Cognitive Science*, 13:259-294.

Dodhiawala, R. T. (1989). Blackboard Systems in Real-Time Problem Solving. In Jagannathan, V., Dodhiawala, R. & Baum, L. S. (eds.), *Blackboard Architectures and Applications*, 181-191. Boston: Academic Press, Inc.

Duncan, S. Jr. (1972). Some Signals and Rules for Taking Speaking Turns in Conversations. *Journal of Personality and Social Psychology*, 23(2), 283-292.

Effron, D. (1941/1972). *Gesture, Race and Culture*. The Hague: Mouton.

Ekman, P. & W. Friesen (1969). The Repertoire of Non-Verbal Behavior: Categories, Origins, Usage, and Coding. *Semiotica*, 1, 49-98.

Goodwin, M. H. & C. Goodwin (1986). Gesture and Coparticipation in the Activity of Searching for a Word. *Semiotica*, 62(1/2), 51-75.

Goodwin, C. (1981). *Conversational Organization: Interaction Between Speakers and Hearers*. New York, NY: Academic Press.

Goodwin, C. (1986). Gestures as a Resource for the Organization of Mutual Orientation. *Semiotica*, 62(1/2), 29-49.

Grice, H. P. (1989). *Studies in the Way of Words*. Cambridge, Massachusetts: Harvard University Press.

Grosz, B. J. & C. L. Sidner (1986). Attention, Intentions, and the Strucutre of Discourse. *Computational Linguistics*, 12(3), 175-204.

Kahneman, D. (1973). *Attention and Effort*. New Jersey: Prentice-Hall, Inc.

Kleinke, C. (1986). Gaze and Eye Contact: A Research Review. *Psychological Bulletin*, 100(1), 78-100.

Kosslyn, S. M. & O. Koenig (1992). *Wet Mind: The New Cognitive Neuroscience*. New York, New York: The Free Press.

Lenat, D. B. (1995). Cyc: A Large-Scale Investment in Knowledge Infrastructure. *Communications of the ACM*, 38(11).

Maes, P. (ed.) (1990a). *Designing Autonomous Agents: Theory and Practice from Biology to Engineering and Back*. Cambridge, MA: MIT Press/Elsevier.

Maes, P. (1990b). Situated Agents can have Goals. In P. Maes (ed.), *Designing Autonomous Agents*, 49-70. Cambridge, MA: MIT Press.

McNeill, D. (1992). *Hand and Mind: What Gestures Reveal about Thought*. Chicago, IL: University of Chicago Press.

Nespolous, J-L & Lecours, A. R. (1986). Gestures: Nature and Function. In J-L Nespolous, P. Perron & A. R. Lecours (eds.), *The Biological Foundations of Gestures: Motor and Semiotic Aspects*, 49-62. Hillsdale, NJ: Lawrence Earlbaum Associates.

Newell, A. (1990). *Unified Theories of Cognition*. Cambridge, MA: Harvard University Press.

Nii, P. (1989). Blackboard Systems. In A. Barr, P. R. Cohen & E. A. Feigenbaum (eds.), *The Handbook of Artificial Intelligence*, Vol. IV, 1-74. Reading, MA: Addison-Wesley Publishing Co.

Pierrehumbert, J. & J. Hirschberg (1990). The Meaning of Intonational Contours in the Interpretation of Discourse. In P. R. Cohen, J. Morgan & M. E. Pollack (eds.), *Intentions in Communication*. Cambridge: MIT Press.

Rimé, B. & Schiaratura, L. (1991). Gesture and Speech. In R. S. Feldman & B. Rimé, *Fundamentals of Nonverbal Behavior*, 239-281. New York: Press Syndicate of the University of Cambridge.

Sacks, H., Schegloff, E. A.. & Jefferson, G. A. (1974). A Simplest Systematics for the Organization of Turn-Taking in Conversation. *Language*, 50, 696-735.

Sacks, H. (1992). *Lectures on Conversation, vol II*. Cambridge, MA: Blackwell.

Schegloff, E. A. & H. Sacks (1973). Opening up Closings. *Semiotica*, 7, 289-327.

Selfridge, O. (1959). Pandemonium: A Paradigm for Learning. *Proceedings of Symposium on the Mechanization of Thought Processes, 1959*, 511-29.

Sommer, R. (1959). Studies in Personal Space. *Sociometry*, 23, 247-260.

Taylor, T. J. & D. Cameron (1987). *Analysing Conversation: Rules and Units in the Structure of Talk*. Oxford, England: Pergamon Press.

Thórisson, K. R. (in press). Machine Perception of Embodied, Real-Time, Multimodal Dialogue. To be published in P. McKevitt (ed.), *Language, Vision and Music*.

Thórisson, K. R. (1999). A Mind Model for Multimodal Communicative Creatures & Humanoids. *International Journal of Applied Artificial Intelligence*, 1999, Vol. 13 (4-5), 449-486.

Thórisson, K. R. (1998). Decision Making in Real-Time Face-to-Face Multimodal Communication. *Second ACM International Conference on Autonomous Agents '98*, Minneapolis, Minnesota, May 12-15.

Thórisson, K. R. (1997). Layered, Modular Action Control in Communicative Humanoids. *Proceedings of Computer Graphics Europe '97*, June 5-7, Genieva, 134-143.

Thórisson, K. R. (1996). Communicative Humanoids: A Computational Model of Psychosocial Dialogue Skills. Ph.D. Thesis, Massachusetts Institute of Technology, U.S.A.

Walker, M. & Whittaker, S. (1990). Mixed Initiative in Dialogue: An Investigation into Discourse Segmentation. *Proceedings of the 28th Annual Meeting of the Association for Computational Linguistics*.

Whittaker, S., S. E. Brennan & H. H. Clark (1991). Co-ordinated Activity: An Analysis of Interaction in Computer-Supported Co-operative Work. *Proceedings of Conference on Computer Human Interaction*, 361-367.

Whittaker, S. & Stenton, P. (1988). Cues and Control in Expert-Client Dialogues. *Proc. 26th Annual Meeting of the Association of Computational Linguistics*, 123-130.

Yngve, V. H. (1970). On Getting a Word in Edgewise. *Papers from the Sixth Regional Meeting.*, Chicago Linguistics Society, 567-78.

7. AFFILIATION

Kris R. Thórisson
Communicative Machines Inc.
131 E 23rd St., suite 2C
New York, NY 10010
http://www.communicativemachines.com

BJÖRN GRANSTRÖM, DAVID HOUSE AND JONAS BESKOW

SPEECH AND GESTURES FOR TALKING FACES IN CONVERSATIONAL DIALOGUE SYSTEMS

1. INTRODUCTION

Innovative spoken dialogue systems are beginning to be characterized by designs where interactivity is no longer seen as limited to a series of choices in a question and answer menu approach. New systems strive toward establishing a smooth flow of information modelled on conversational dialogues. In this context, there is currently considerable interest in developing 3D-animated agents to exploit the inherently multimodal nature of speech communication. As the 3D-animation becomes more sophisticated in terms of visual realism, the demand for naturalness in speech and gesture coordination increases. Not only are appropriate and speech-synchronized articulator movements necessary, conversational signals such as cues for turntaking and feedback are also essential. Such conversational signals can be conveyed by both the auditory and visual modality. Verbal (auditory) signals can complement syntax and interact with the prosodic (accentual and phrasal) structure of the utterances. For example, a phrase-final intonation pattern can function as both a cue for prosodic grouping and as a verbal turngiving signal. Gestural (visual) signals such as eyebrow movements and nodding for accentuation can function as parallel signals to intonation (i.e. as linguistic signals) as well as being used as conversational signals (e.g. raised eyebrows to signify an interested, listening agent or nodding to provide encouragement). While much work has been done on describing spoken and gestural conversational signals in human-to-human interaction (see e.g. Allwood, this volume; McNeill, this volume), work aimed at investigating the coordination of these two types of signals in computer-human interaction and the implementation of this knowledge in animated conversational agents is still relatively scarce.

Cassel et al. (1994) have modelled speech and gesture in dialogue using two virtual agents, but no user interactivity. Katashi & Akikazu (1994) employed animated facial expressions, but no gestures, as a back-channelling mechanism in a spoken dialogue system. Thórisson (1999, and this volume) used an animated character together with input from many sources, including speech and gaze, to model mainly the social aspects of multi-modal dialogue interaction. A good summary of relatively recent work in this area can be found in Cassell et al. (2000a).

This chapter presents a background and description of an audio-visual synthesis system developed at KTH, experimental results using this system and finally some

B. Granström et al. (eds.), Multimodality in Language and Speech Systems, 209–241.

examples of experimental dialogue systems which make use of the speech and gesture potentials of audio-visual synthesis.

2. MODELLING AND SYNTHESIS OF TALKING FACES

2.1. Facial Animation Methods

There exist several approaches to facial animation, ranging from display of pre-stored images to actual physical simulation. Platt & Badler (1981), Waters (1987) and Terzopoulos & Waters (1990) have all developed models that simulate the muscles and tensions in the facial tissue. Such models can provide realistic results, but determining proper muscle activation levels required to create a specific facial expression can be a difficult matter, since real muscles do not lend themselves to easy measurement. Furthermore, muscle-based models tend to be computationally expensive, which is a serious drawback if the intended application is e.g. an inter-active interface agent. More suitable in this respect is the direct parameterisation approach taken by Parke (1982). Here, attention is turned solely to the outside surface of the face, and instead of muscles, a set of observation-based parameters is used to deform the surface to create facial expressions. Such surface deformations can often be of low computational complexity. Since the parameters are not required to model any real anatomical features, they can be tailored to mimic particular actions, such as speech movements. Further-more, parameters can relatively easily be measured from real faces, either manually or using image processing techniques (Öhman, 1998). Additional levels of naturalness such as dynamic facial wrinkles and furrows have also been implemented at the surface level (avoiding computationally expensive physical modelling) by taking advantage of recent developments in graphics hardware such as the ability to do bump-mapping (Pelachaud, et al. 2001).

A common drawback of direct parameterisation methods as well as some muscle-based schemes, however, is that activation of several parameters simul-taneously can yield unpredictable results. Consider a scheme with independent control parameters for the upper and lower lip, as well as one parameter for jaw opening. In order to close the lips tightly together, we need to specify an exact combination of all three parameters. It is possible to solve this kind of problem using collision detection and iterative parameter adjustment, but this dramatically increases the computational demand.

In analogy with the techniques for concatenating natural speech, used for auditory synthesis, several schemes for visual speech, using image processing techniques have been used with good results (Ezzat & Tomaso, 1998; Bregler, Covell & Slaney, 1997; Brooke & Scott, 1998). Similar to the case in auditory concatenative synthesis the naturalness is often very high, but the techniques lack some of the flexibility of parametric synthesis.

2.2. Audio-Visual Synthesis at KTH

The audio-visual text-to-speech synthesis system is based on the KTH rule-based text-to-speech synthesis, employing either formant synthesis (Carlson, Granström & Hunnicutt, 1991) or MBROLA concatenation for acoustic signal generation. The visual signal generation is based on 3D polygon models, which are parametrically articulated and deformed.

The facial models are parameterised using weighted geometric transformations (translation, scaling and rotation) of the vertices. This is a generalised version of the scheme proposed by Parke (1982); see Beskow (1997) for details. For each deformation, weights are set individually for each vertex in the model. For example, for jaw rotation, the transformation in question is a rotation around the jaw hinge, and the vertices on the lower lip and chin are given weights close to 1.0 while vertices on the side of the mandible and lower cheeks are given weights closer to zero. This simulates the elasticity of the skin. Furthermore, the transformations are normalised with respect to target areas in the structure, in such a way that full application of a deformation parameter ensures that the articulator in question reaches a well-defined target. An example is labiodental occlusion, where the articulator (the lower lip) will reach the target (the lower edge of the upper front teeth) when the deformation parameter is fully applied.

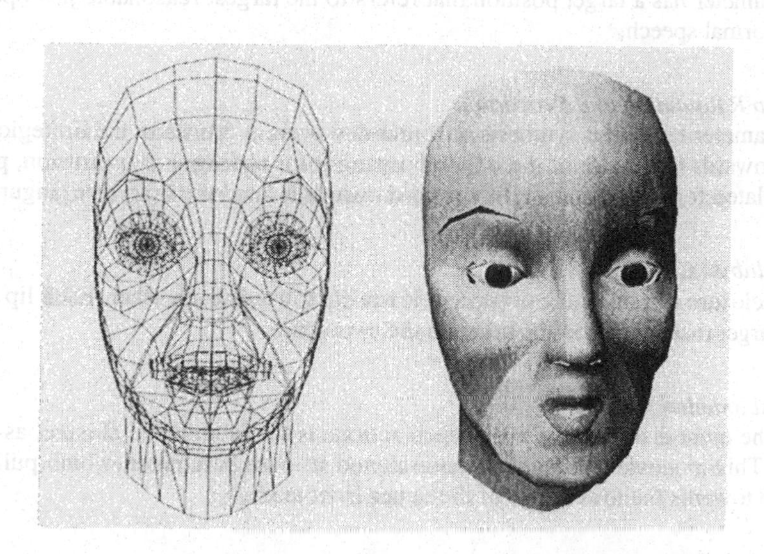

Figure 1. The original KTH face-synthesis model, based on Parke (1982).

2.2.1. Articulators

The parameter set used for speech animation is chosen to reflect the articulatory phonetic features often used to describe speech production, and most parameters are defined in terms of articulatory targets rather than general geometric measures (see Table 1). This makes it easy to develop the rules to control the system. The most important articulators and articulatory parameters are described in the following sections.

Table 1. Parameters used for articulatory control of the face.

Parameters
Jaw rotation
Lip rounding
Lip protrusion
Mouth width
Bilabial closure
Labiodental closure
Tongue tip

2.2.2. Jaw Rotation

Jaw rotation simply rotates all vertices related to the mandible around the jaw hinge. This parameter has a target position that refers to the largest reasonable jaw opening during normal speech.

2.2.3. Lip Rounding and Protrusion

This parameter facilitates synthesis of rounded vowels. Vertices in the lip region are moved towards the centre of the mouth opening for rounding. For protrusion, points are translated forward and slightly rotated outwards around the lip centre (Figure 2).

2.2.4. Bilabial Closure

Bilabial closure translates the upper and lower lip towards each other. Each lip has a virtual target that is the midline between the two lips.

2.2.5. Labiodental Closure

One of the most distinctive visual speech actions is the labiodental closure, as in /f/ and /v/. This movement has also been assigned its own parameter, which pulls the lower lip towards the lower edge of the upper front teeth.

2.2.6. Tongue

Human speech production relies heavily on tongue movements. Tongue actions are important not only auditorily but also visually. Visibility of the tongue during speech is speaker dependent, due to for example individual articulation style.

Clearly visible tongue movements can, however, raise the intelligibility of speech. In order to allow synthesis of visual speech without omitting any potentially important visual information, a tongue was modelled and added to the talking head.

In natural speech, the apically articulated phonemes are responsible for most visually perceivable tongue actions. Thus, for the purposes of visual speech, it has proven adequate to stylise the tongue as a simple one parameter "flap". The motion of the tongue tip is normalised with respect to the alveolar ridge (Figure 3). It should be noted that for certain applications such as language tutoring, it is desirable with a more detailed and realistic tongue model. Such a model is currently being developed at the department (Engwall, 2001).

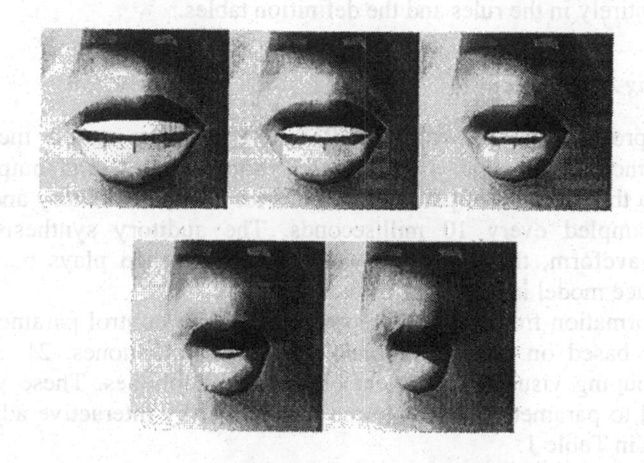

Figure 2. Lip shape for different values of the lip rounding parameter: 0, 25, 50, 75 and 100 % rounding. Jaw rotation is 5 degrees.

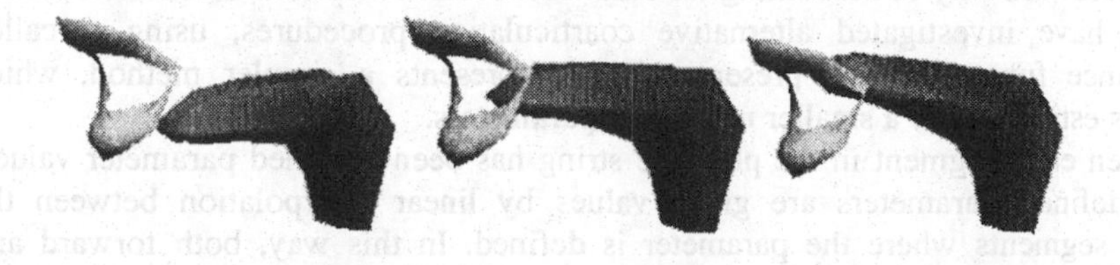

Figure 3. View of lips and tongue with different tongue tip values: 0; 50; 100% raising with reference to the alveolar ridge.

2.3. System Description

The audiovisual text-to-speech system consists of three parts:

- The auditory speech signal generator, that can be either formant based or concatenation based.
- The visual signal generator, i.e. a facial model as described above.
- The rule synthesis framework, RULSYS. This is a generic rule synthesis system that transforms input text (orthographic or phonetic) to synthesiser control parameters. The system itself is independent of both language and synthesiser used to produce output. The language-specific and synthesiser-specific functionality lies entirely in the rules and the definition tables.

2.4. Controlling Articulation

In brief, the process works as follows: RULSYS operates on the incoming text according to the morphologic, syntactic and phonetic rules, and outputs control parameter data that consists of parameter values for both the auditory and the visual synthesiser, sampled every 10 milliseconds. The auditory synthesiser module generates a waveform, that the face synthesis module then plays back, while it animates the face model in synchrony.

The transformation from phonetic representation into control parameters for the face model is based on visemes. From 45 Swedish allophones, 21 visemes are created by grouping visually equivalent or similar allophones. These visemes are then translated to parameter settings in the face model by interactive adjustment of the parameters in Table 1.

For each viseme, following the general principle of the acoustic synthesis, each of the parameters is either assigned a value, or is left undefined. If a parameter is left undefined, it simply means that the viseme is independent of that particular parameter. (For example, /r/ can be either rounded or unrounded, depending on the context, thus the lip rounding parameter is left undefined for this viseme.) This is a straightforward way of accounting for basic coarticulation effects. Cohen & Massaro (1993) have investigated alternative coarticulation procedures, using so-called dominance functions. The present scheme represents a simpler method, which requires estimation of a smaller number of parameters.

When each segment in the phonetic string has been assigned parameter values, the undefined parameters are given values by linear interpolation between the nearest segments where the parameter is defined. In this way, both forward and backward coarticulation is modelled.

In RULSYS, the output parameters can be filtered using different time variable filters in order to obtain realistic dynamic properties. The face synthesis parameters are assigned different time constants depending on e.g. the mass of the corresponding speech organ.

2.5. Visual Prosody

The visible articulatory movements are mainly those of the lips, jaw and tongue. However, these are not the only visual information carriers in the face during speech. Much information related to phrasing, stress, intonation and emotion are expressed by for example nodding of the head, raising and shaping of the eyebrows, eye movements and blinks.

These kinds of facial actions should also be taken into account in a visual speech synthesis system, not only because they may transmit important non-verbal information, but also because they make the face look alive.

These movements are more difficult to model in a general way than the articulatory movements, since they are optional and highly dependent on the speaker's personality, mood, and purpose of the utterance, etc.

Nevertheless, a few general rules apply to most speakers. For example, it is quite common to raise the eyebrows at the end of a question and to raise the eyebrows or nod the head during a stressed syllable. We have developed an advance research tool, which is used to experiment with prosodic signals on both a parametric and symbolic level. The tool is described in more detail in the following sections.

2.6. Parameter Manipulation

For stimuli preparation and explorative investigations, we have developed a control interface that allows fine-grained control over the trajectories for acoustic as well as visual parameters. The interface is implemented as an extension to the WaveSurfer application (Beskow & Sjölander, 2000), which is a tool for recording, playing, editing, viewing, printing, and labelling audio data. The interface makes it possible to start with an utterance synthesised from text, with all the parameters generated by rule, and then interactively edit the parameter tracks for any parameter, including voice fundamental frequency (F0), visual (non-articulatory) parameters as well as the durations of individual segments in the utterance to produce specific effects.

An example of the user interface is shown in Figure 4. In the top box, a text can be entered in Swedish or English. This creates a phonetic transcription that can then be edited. On pushing "Synthesize", rule generated parameters will be created and displayed in different panes below. The selection of parameters is user controlled. The lower section contains segmentation and the acoustic waveform. The talking face is displayed in a separate window.

The acoustic synthesis can be exchanged for a natural utterance and synchronised to the face synthesis. This is useful for different experiments on multimodal integration and has been used in the Teleface project (Agelfors et al., 1999), aiming at a telephone device for hard-of-hearing persons, as described in Section 3. In automatic language learning and pronunciation training applications it can be used to add to the naturalness of the tutor's voice in cases when the acoustic synthesis is judged to be inappropriate.

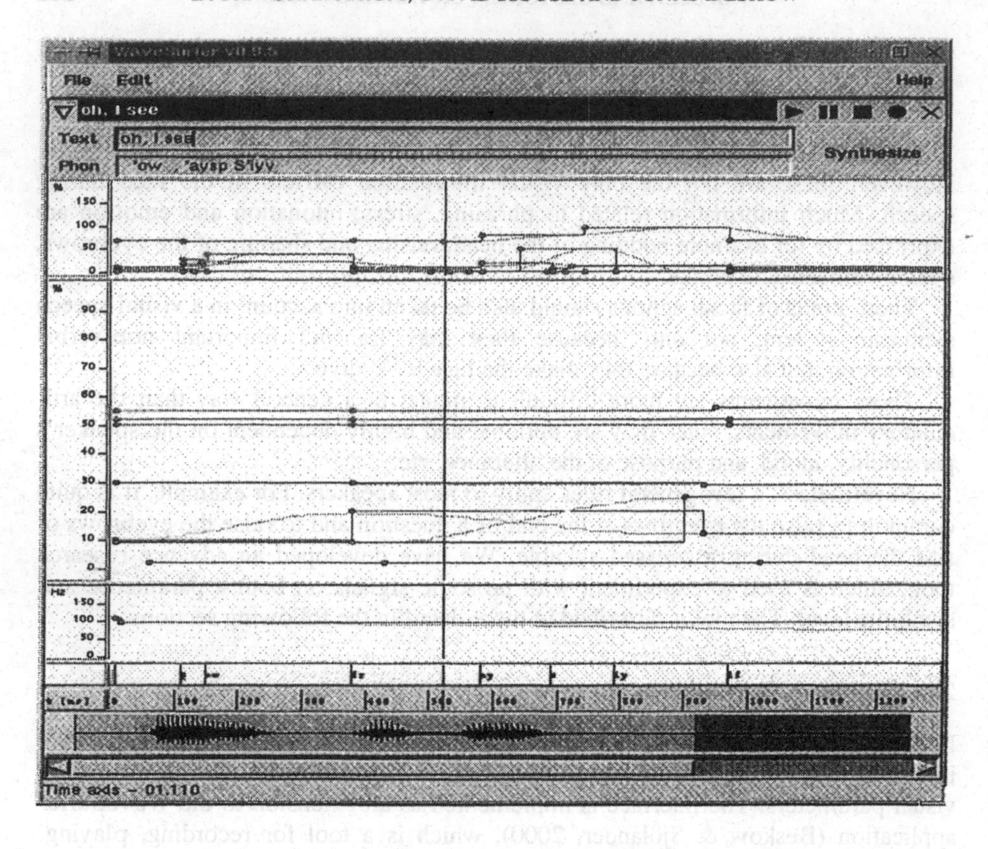

Figure 4. The WaveSurfer user interface for parametric manipulation of the multimodal synthesis.

2.7 Symbolic Representation

The parametric manipulation tool described in the previous section is used to experiment with and define different gestures. A gesture library is under construction, containing primitives for general emotions and non-speech specific gestures as well as some primitives for linguistic cues. These primitives serve as a base for the creation of new communicative gestures in future animated talking agents used in multimodal spoken dialogue systems and as automatic tutors.

To enable display of the agents' different moods, basic emotions based on the six universal emotions defined by Ekman (1979) have been implemented in a way similar to that described by Pelachaud, Badler & Steedman (1996).

We are at present developing an XML-based representation of visual cues that facilitates description of the cues at a higher level. It is important that there is a one-to-many relation between the symbols and the actual gesture implementation to avoid stereotypic agent behaviour.

2.8. Different Speakers

The basic techniques for articulatory deformation described in Section 2.1 can be generalised to any 3D wireframe model of a face. We have designed a procedure to interactively define deformation parameters and their influence on the wireframe using a graphical editor (Beskow, 1997). Using this technique, we can relatively easily extend our gallery of talking faces. To date, a large number of characters with different characteristics have been designed. Four examples are shown in Figure 5.

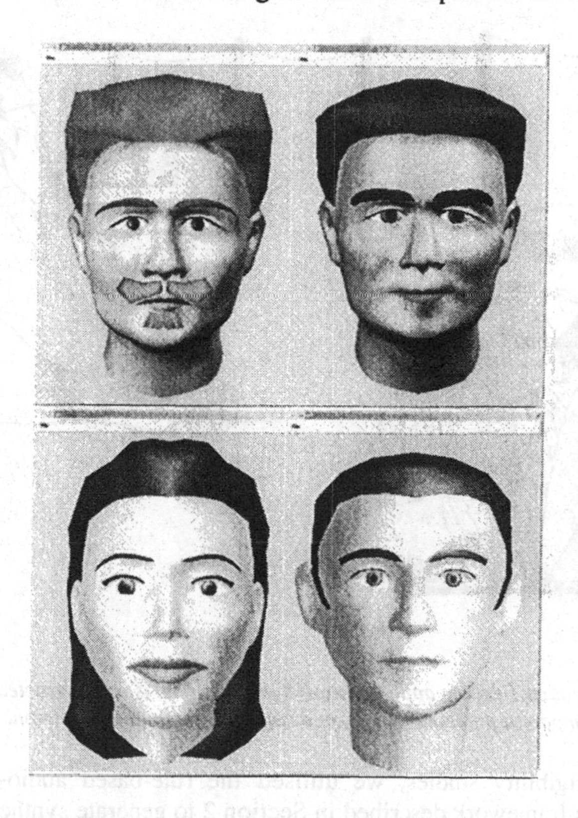

Figure 5. Examples of different talking faces created for use in conversational systems.

All faces are parameterised using the same set of basic speech parameters (Table 1), which has the distinct advantage that the control parameters generated by the text-to-speech system can interchangeably control any of the several models during speech.

3. MULTI-MODAL SPEECH INTELLIGIBILITY – THE TELEFACE PROJECT

The speech intelligibility of talking animated agents, as the ones described above, has been tested within the Teleface project at KTH (Beskow et al., 1997; Agelfors et al., 1998). The project focuses on the usage of multi-modal speech technology for hearing-impaired persons. The aim of the first phase of the project was to evaluate the increased intelligibility hearing-impaired persons experience from an auditory signal when it is complemented by a synthesised face. In this case, techniques for combining natural speech with lip-synchronised face synthesis have been developed. A demonstrator of a system for telephony with a synthetic face that articulates in synchrony with a natural voice is currently being implemented (see Figure 6).

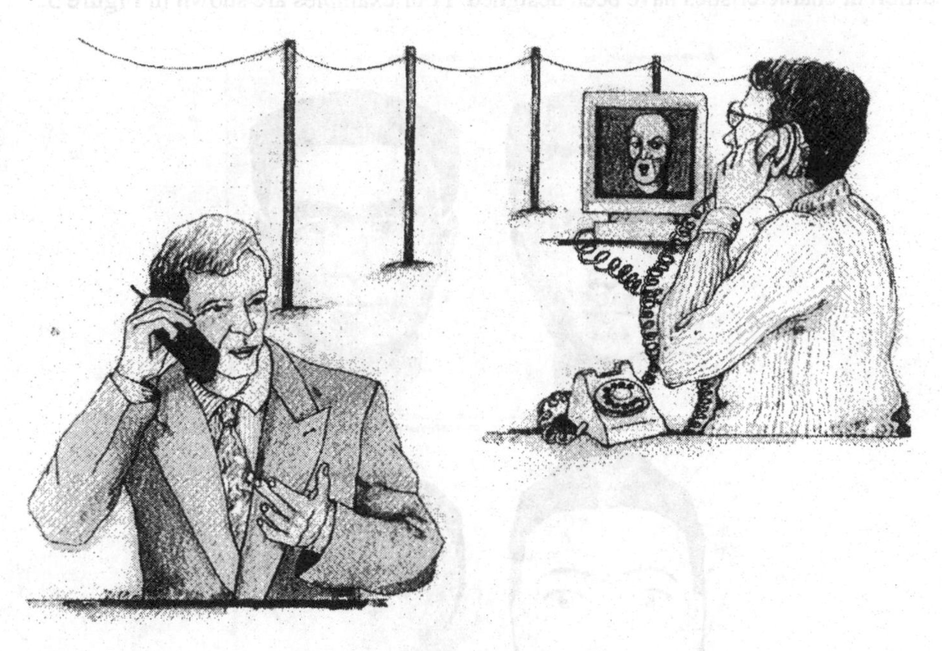

Figure 6. The intended Teleface application. A synthetic face is reconstructed at the hearing-impaired person's end of an ordinary telephone connection.

In the intelligibility studies, we utilised the rule-based audio-visual text-to-speech synthesis framework described in Section 2 to generate synthetic acoustic, as well as visual, speech stimuli. Automatic extraction of facial parameters from the acoustic signal requires extensive analysis of the relationship between the facial parameters and the acoustics. To this end, we have built a framework for automatic measurements of visible speech movements (Öhman, 1998). A database of video sequences of a male speaker has been recorded. The speaker is a Swedish male from Stockholm. Parts of the speaker's face have been marked with a blue colour to facilitate image analysis of lips and other parts of the face that are important for speech reading. The parameters from the optical measurements were statistically analysed together with the acoustic signal, providing knowledge about the relationship between the visual and acoustic modes of speech. Thus optical measurements were used to improve naturalness of the visual speech synthesis.

3.1. Intelligibility Test with Normal-Hearing Subjects

A database of video recordings of a male speaker's face pronouncing Swedish VCV-words and everyday sentences was recorded as a basis for the intelligibility test. The audio track was separated from the video recordings and phonetically labelled. By processing the label file through the rule system for audio-visual text-to-speech synthesis, parameter trajectories for face-model animation as well as formant-synthesiser control were calculated. In the synthesis procedure, all the phoneme durations were copied from the original utterances in the video sequences. The parameter files were used to generate two different synthetic voices as well as animations of two different synthetic faces, using the original KTH multi-modal synthesis and the cartoon like Olga version described in Section 5.2. With the natural face and the natural voice, this adds up to three faces and three voices, where the audio and animation are synchronised in each case.

The subjects were 18 fourth-year MSc-students at KTH. The test was carried out as part of a mandatory lab exercise in the Speech Communication course given by the department. A screening test was performed to check that all the subjects had normal hearing.

In the first test series, we used lists consisting of VCV-words with 17 Swedish consonants, /b,d,g,p,t,k,s,ʃ,ç,f,v,m,n,ŋ,j,l,r/, in symmetric context with the vowels /a/ and /ʊ/. When performing tests with this material on normal-hearing persons, the audio signal was degraded by adding white noise. The signal-to-noise ratio in these tests was 3 dB.

The tests were performed in a computer-based test environment (Lundeberg, 1997). This gave us the opportunity to play video sequences of the faces with sound files of the voices. In this way, it was possible to evaluate the intelligibility of different audio-visual combinations. A monitor was used for presenting the visual stimuli and a loudspeaker for the audio. A forced choice response for the VCV-words was made using the mouse on the computer screen presenting all consonants in the stimuli set. There was no time limit for the response.

3.2. Results

Data from the tests were analysed using confusion matrices and feature analysis. Overall results are shown in Figure 7. The differences between the two synthetic versions were not significant. Adding a synthetic face to a natural male voice improved the correct response rates from 63% to 70%. The corresponding result when adding a natural face was 76%. Synthetic male voice gave 31% correct responses compared to 45% with a synthetic face added. Results for the consonants in the context of /a/ were generally better than in the context of /ʊ/, both acoustically and visually.

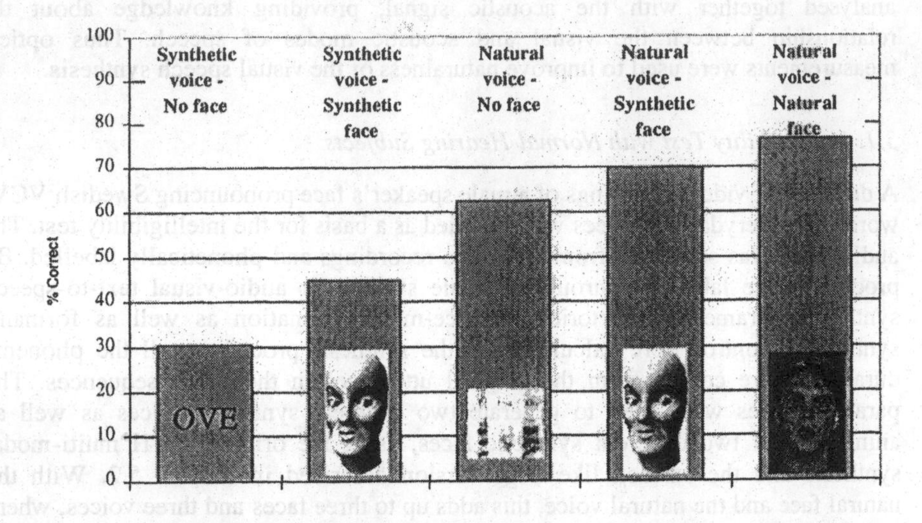

Figure 7. Results from intelligibility tests. Number of correctly identified consonants in VCV words (in %).

The proportion of correct responses with respect to place of articulation ('total' in Figure 8), increased from 72% for natural voice only, to 83% with a synthetic face and to 86% with a natural face. The highest improvement when adding a synthetic or natural face is observed for bilabials and labiodentals. This is hardly surprising, since they are two of the most salient visemes. It is also interesting to note that these categories have poor voice-only scores.

Dentals showed almost no difference between the audio and audio-visual conditions. Palatal and velar consonants did not benefit at all from adding a face to the natural voice condition. This is not surprising considering the high voice-only score in combination with the back articulatory movements. The results give a good illustration of the complementary nature of the visual and auditory information in multimodal speech perception. The visual mode is most important for the front consonants (bilabials and labiodentals received low auditory but high audio-visual score), whereas acoustic-only information is most important for back consonants. Results show that a synthetic face substantially improves the intelligibility of synthetic and natural speech.

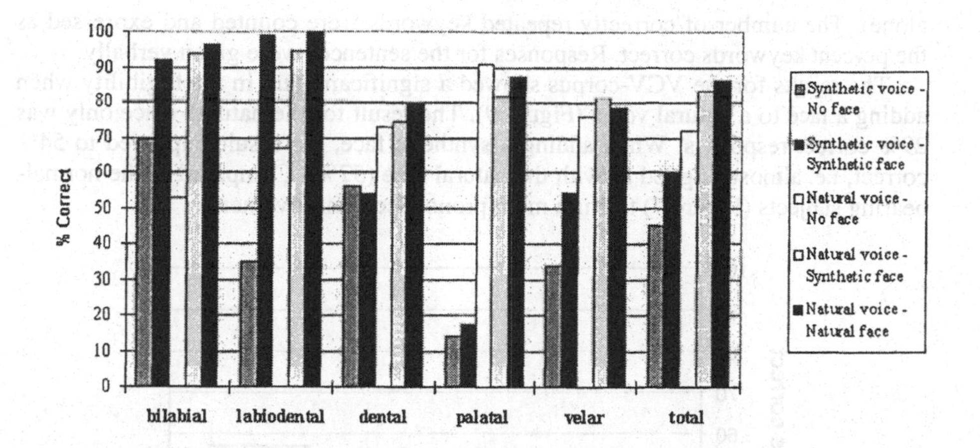

Figure 8. Results from feature analysis of intelligibility tests. Correct responses (in %) with respect to five places of articulation: bilabials /p, b, m/, labiodentals /f, v/, dentals /t, d, s, l, r, n/, palatals /j, ç, ʃ/, and velars /k, g, ŋ/. The overall result with respect to place of articulation is shown as 'total'.

3.3. Speech Intelligibility Tests with Hearing-Impaired Subjects

The ultimate test of the usefulness of synthesised faces in the Teleface application is the performance of hearing-impaired subjects. We have replicated the VCV-test described in the previous section and also used the more realistic sentence material from our audio-visual database.

We tested 12 subjects with a severe to profound hearing loss for the better ear with a mean of 88.4 dB HL pure tone average hearing threshold loss, (ranging from 62-103 dB HL) at the frequencies 0.5, 1 and 2 kHz. The subjects were between 23 and 76 years old (mean: 58.2 years). The speech was presented at about 70 dB SPL and each subject was allowed to adjust their hearing aid to their most comfortable listening level during a training session (five sentences each of the natural and the synthetic faces).

The test material consisted of VCV-syllables and "everyday sentences". The VCV-test was similar to the one described in the previous section. The VCV-syllables consisted of seventeen consonants presented in /aCa/, /ɪCɪ/, /ʊCʊ/ context (C={b,d,g,p,t,k,s,ʃ,ç,f,v,m,n,ŋ,j,l,r}). Three test lists (3 x 17 stimuli/list) were presented in different conditions for the subjects (two audio-visual and one audio-alone). Subjects were asked to respond with the consonant. Responses for the VCV-corpus were forced-choice and made with a graphical interface on a computer screen.

The sentences were unrelated Swedish "everyday sentences" previously developed at our department, based on MacLeod & Summerfield (1990). Each list contained 15 sentences (three keywords/sentence, i.e. 45 keywords). Six test lists were presented, two test lists in each condition (two audio-visual and one audio alone). The number of correctly repeated keywords were counted and expressed as the percent keywords correct. Responses for the sentences were given verbally.

The results for the VCV-corpus showed a significant gain in intelligibility when adding a face to a natural voice (Figure 9). The result for the natural-voice only was 29% correct responses. When adding a synthetic face, the result improved to 54% correct, i.e. almost as good as with the natural face (57%). Compared to the normal-hearing subjects (Figure 7) this is a more pronounced improvement.

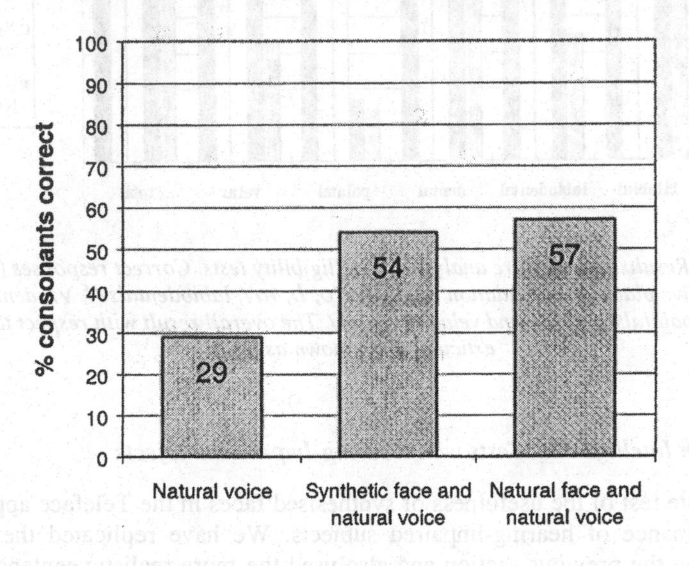

Figure 9. Results for hearing-impaired subjects on the VCV test.

The difference between natural and synthetic faces seems to influence the result less for the hearing-impaired persons. A possible explanation is a general superior speech reading proficiency in the hearing-impaired group. There was, however, large individual variability.

The mean results for the sentences gave 41% keyword correct in the audio-only case while the scores for the audio plus synthetic or natural face were 65% and 82%, respectively. This is consistent with the VCV-results and further demonstrates the importance of multimodality in speech perception.

4. PHRASING AND PROMINENCE PERCEPTION EXPERIMENTS

The interaction between acoustic and visual cues in communication has been discussed previously. Specifically the interaction between acoustic intonational gestures (F0) and eyebrow movements has been studied in e .g. Cavé et al. (1996). We performed an informal experiment where F0 and eyebrow movements were directly coupled and concluded that such a direct relation is very unnatural, but it is nonetheless clear that eyebrow movements may co-occur with accented syllables.

To investigate this more formally the following pair of experiments was designed to test the independent contribution of eyebrow movements to the audio-visual perception of phrasing and prominence in a single test sentence. The test sentence used to create the stimuli for the experiments was ambiguous in terms of an internal phrase boundary. Acoustic cues and lower face visual cues were the same for all stimuli. Articulatory movements were created by using the text-to-speech rule system. The upper face cues were eyebrow movements where the eyebrows were raised on successive words in the sentence.

The movements were created by hand editing the eyebrow parameter using the synthesis parameter editor. The degree of eyebrow raising was chosen to create a subtle movement that was distinctive although not too obvious. The total duration of the movement was 500 ms and comprised a 100 ms dynamic raising part, a 200 ms static raised portion and a 200 ms dynamic lowering part. The synthetic face *Alf* with neutral and with raised eyebrows is shown in Figure 10.

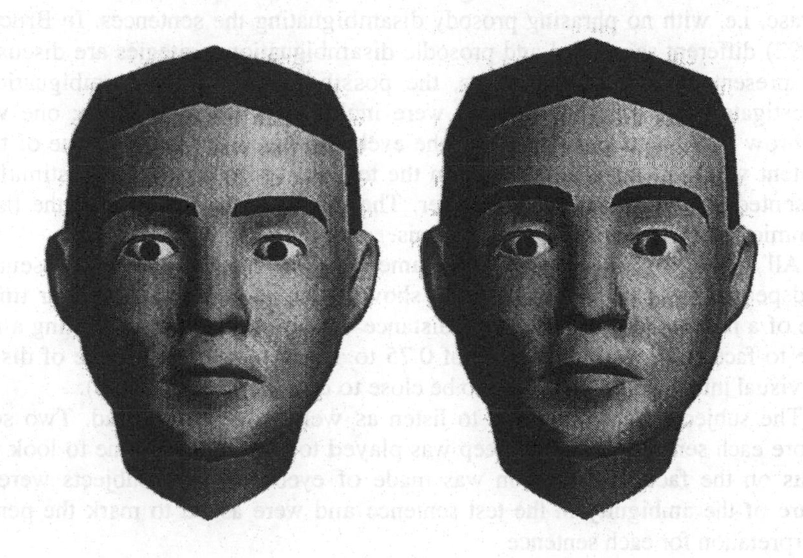

Figure 10. The synthetic face Alf *with neutral eyebrows (left) and with eyebrows raised (right).*

In the two experiments described below the same 21 subjects participated. All were part of a speech technology class taught at KTH. No one reported any hearing loss or visual impairment. 14 subjects had Swedish as their mother tongue. All except one of these reported that they had a central Swedish (Stockholm) dialect.

Seven subjects had other mother tongues than Swedish (1 Finnish, 2 French, 2 Italian and 2 Spanish), but all had working competence in Swedish (attending a master's level class given in Swedish at KTH). Below, the results of the total group as well as for these subgroups are presented.

4.1. Experiment 1 – Phrasing

In a previous study concerned with prominence and phrasing, using acoustic speech only, ambiguous sentences were used (Bruce et al., 1992). In the present experiment, we used one of these sentences:

(1) När pappa fiskar stör, piper Putte
(When dad is fishing sturgeon, Putte is whimpering)
(2) När pappa fiskar, stör Piper Putte
(When dad is fishing, Piper disturbs Putte).

Hence, "stör" could be interpreted as either a noun (1) (sturgeon) or a verb (2) (disturbs); "piper" (1) is a verb (whimpering), while "Piper" (2) is a name.

In the stimuli, the acoustic signal is always the same, and synthesized as one phrase, i.e. with no phrasing prosody disambiguating the sentences. In Bruce et al. (1992) different segmental and prosodic disambiguation strategies are discussed. In the present series of experiments, the possibility of visual disambiguation was investigated. Six different versions were included in the experiment: one with no eyebrow movement and five where the eyebrow rise was placed on one of the five content words in the test sentence. In the test list of 20 stimuli, each stimulus was presented three times in random order. The first and the last item of the list were dummies and not part of the data analysis.

All subjects participated in the same session. The audio was presented via loudspeakers and the face image was shown on a projected screen, four times the size of a normal head. The viewing distance was 3 to 6 meters, simulating a normal face-to-face conversation distance of 0.75 to 1.5 meters. In this range of distances, the visual intelligibility is judged to be close to constant (Neely, 1956).

The subjects were instructed to listen as well as to speech read. Two seconds before each sentence an audio beep was played to give subjects time to look up and focus on the face. No mention was made of eyebrows. The subjects were made aware of the ambiguity in the test sentence and were asked to mark the perceived interpretation for each sentence.

In Figure 11 the results from experiment 1 can be seen. It is obvious that there is a bias for all the stimuli to more often (about 60%) be perceived with a phrase boundary after "stör", i.e. interpretation (1).

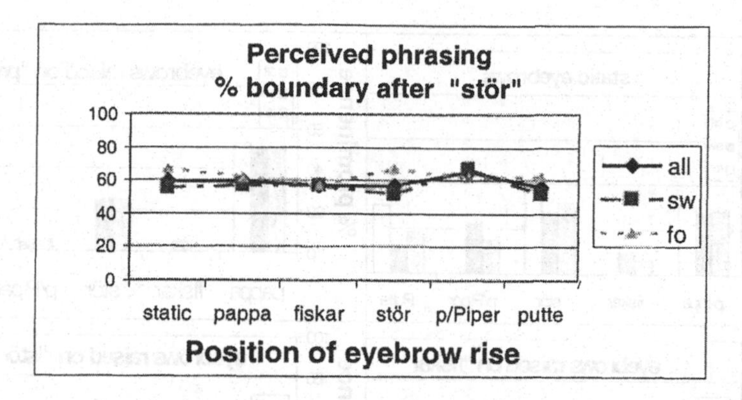

Figure 11. Result of the phrasing/disambiguation experiment. Interpretation (1), i.e. a phrase boundary after "stör" (rather than after "fiskar"). sw: 14 subjects with Swedish as their mother tongue. fo: *7 non-Swedish subjects*

The non-Swedish subjects behaved very much like the Swedes, perhaps with one exception. For the Swedish subjects, there was a small increase in the (1) interpretation when there was an eyebrow rise on p/Piper. One possible explanation could be that an eyebrow movement could be associated with a phrase onset, but on the whole there is rather limited evidence in this experiment that eyebrow movements contributed to phrasing information.

4.2. Experiment 2 – Prominence

In the second experiment, we used the same stimulus material as in experiment 1, but the question was now concerned with prominence. The subjects were asked to circle the word that they perceived as most stressed/most prominent in the sentence.

The results are shown in Figure 12. Figure 12/static refers to judgements when there is no eyebrow movement at all. Obviously the distribution of judgements varies with both subject group and word in the sentence. This could be related to phonetic information in the auditory modality since the intonational default synthesis used here put a weak focal accent on the first and the last word in a sentence. This could explain the many votes for the first and the last word, "pappa" and "Putte" in Figure 12/static. However, it may well be related to prominence expectations. In experiments where subjects are asked to rate prominence on words in written sentences, nouns tend to get higher ratings than verbs (Fant & Kruckenberg, 1989). This is supported by our data, since "stör" has the default interpretation of a noun and p/Piper the default interpretation of a verb in experiment 1, while "fiskar" is always a verb in these contexts. The non-Swedish subjects seem to behave slightly differently in this experiment, since no prominence votes are given to "fiskar" and "p/Piper".

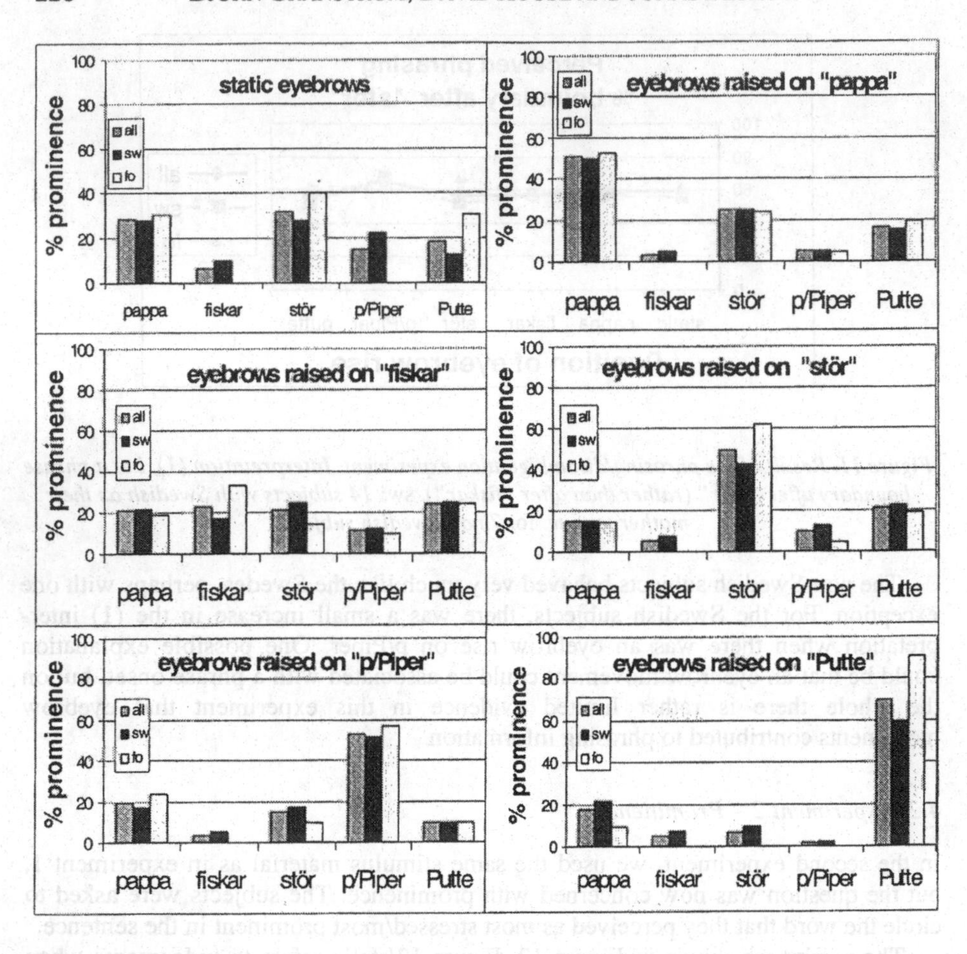

Figure 12. Prominence responses in percent for each word and each stimulus condition. Subjects are grouped as all, Swedish (sw) and foreign (fo).

The results of the prominence experiment indicate that eyebrow raising can function as a perceptual cue to word prominence independent of acoustic cues and lower face visual cues. While there was no systematic manipulation of the acoustic cues in this experiment, a certain interplay between the acoustic and visual cues can be inferred from the results. As mentioned above, a weak acoustic focal accent in the default synthesis falls on the final word "Putte". Eyebrow raising on this word (Figure 12/Putte) produces the greatest prominence response in both listener groups. This could be a cumulative effect of both acoustic and visual cues, although compared to the results where the eyebrows were raised on the other nouns, this effect is not great.

In an integrative model of visual speech perception (Massaro, 1998), eyebrow raising may signal prominence when there is no direct conflict with acoustic cues. In the case of "fiskar" (Figures 12/static and 12/fiskar), the lack of specific acoustic cues for focus and the linguistic bias between nouns and verbs, as mentioned above, could account for the scarcity of prominence responses for "fiskar". Further experimentation where strong acoustic focal accents are coupled to and paired against eyebrow movement could provide more data on this subject.

It is interesting to note that the foreign subjects in all cases responded more consistently to the eyebrow cues for prominence, as can be seen in Figure13. This might be due to the relatively complex stress/tone/focus signalling in terms of F0 in Swedish and the subjects' non-native competence. It could be speculated that the eyebrow motion is a more universal cue for prominence.

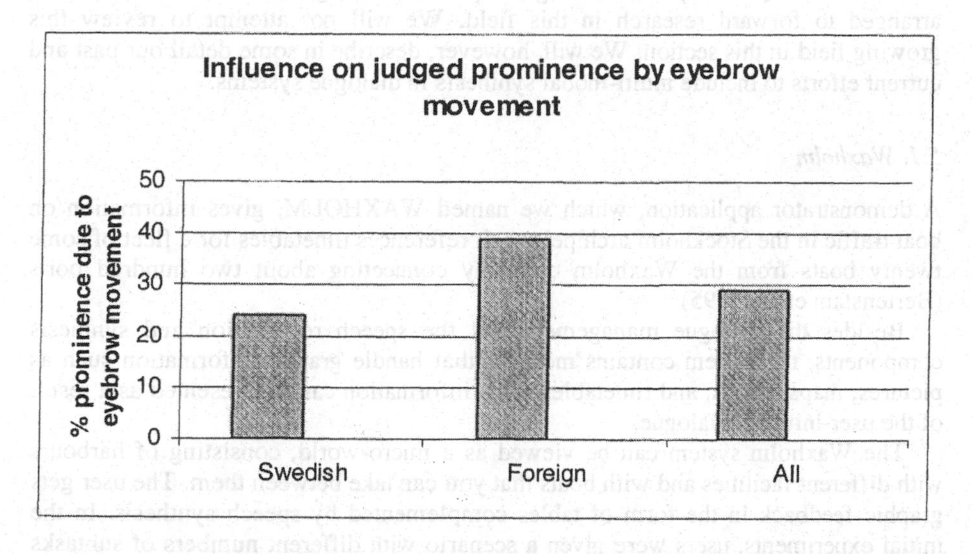

Figure13. Mean increase in prominence judgement due to eyebrow movement.

The relationship between cues for prominence and phrase boundaries is not unproblematic (Bruce et al., 1992). The use of eyebrow movement to signal phrasing may involve more complex movement related to coherence within a phrase rather than simply as a phrase delimiter. It may also be the case that eyebrow raising is not an effective independent cue for phrasing, perhaps because of the more complex nature of different phrasing cues.

This experiment presents evidence that eyebrow movement can serve as an independent cue to prominence. Some interplay between visual and acoustic cues to prominence and between visual cues and word class/prominence expectation are also seen in the results. Eyebrow raising as a cue to phrase boundaries was not shown to be effective as an independent cue in the context of the ambiguous sentence. Further work on the interplay between eyebrow raising as a cue to prominence and eyebrow movement as a visual signal of speaker expression, mood and attitude will benefit the development of interactive animated agents in spoken dialogue systems such as those described in the following section.

5. EXPERIMENTAL DIALOGUE SYSTEMS USING SPEECH AND GESTURE

Our research group at KTH has implemented several systems in which speech synthesis and speech recognition can be studied and developed in a man-machine dialogue framework. Spoken dialogue systems have attracted considerable interest during the past few years. Special workshops and symposia, starting with the 1995 ESCA workshop on Spoken Dialogue Systems in Vigsø, Denmark, have been arranged to forward research in this field. We will not attempt to review this growing field in this section. We will, however, describe in some detail our past and current efforts to include multi-modal synthesis in dialogue systems.

5.1. Waxholm

A demonstrator application, which we named WAXHOLM, gives information on boat traffic in the Stockholm archipelago. It references timetables for a fleet of some twenty boats from the Waxholm company connecting about two hundred ports (Bertenstam et al., 1995)

Besides the dialogue management and the speech recognition and synthesis components, the system contains modules that handle graphic information such as pictures, maps, charts, and timetables. This information can be presented as a result of the user-initiated dialogue.

The Waxholm system can be viewed as a micro-world, consisting of harbours with different facilities and with boats that you can take between them. The user gets graphic feedback in the form of tables complemented by speech synthesis. In the initial experiments, users were given a scenario with different numbers of subtasks to solve. A problem with this approach is that the users tend to use the same vocabulary as the text in the given scenario. We also observed that the user often did not get enough feedback to be able to decide if the system had the same interpretation of the dialogue as the user.

To deal with these problems a graphical representation that visualises the Waxholm micro-world was implemented. An example is shown in Figure 14. One purpose of this was to give the subject an idea of what can be done with the system, without expressing it in words. The interface continuously feeds back the information that the system has obtained from the parsing of the subject's utterance, such as time, departure port and so on. The interface is also meant to give a graphical view of the knowledge the subject has secured thus far, in the form of listings of hotels and so on.

The visual animated talking agent is an integral part of the system. This is expected to raise the intelligibility of the system's responses and questions. Furthermore, the addition of the face into the dialogue system has many other exciting implications. Facial non-verbal signals can be used to support turntaking in the dialogue, and to direct the user's attention in certain ways, e.g. by letting the head turn towards time tables, charts, etc. that appear on the screen during the dialogue. The dialogue system also provides an ideal framework for experiments with non-verbal communication and facial actions at the prosodic level, as discussed above, since the system has a much better knowledge of the discourse context than is the case in plain text-to-speech synthesis.

To make the face more alive, one does not necessarily have to synthesise meaningful non-verbal facial actions. By introducing semi-random eye blinks and very faint eye and head movements, the face looks much more active, and becomes more pleasant to watch. This is especially important when the face is not talking.

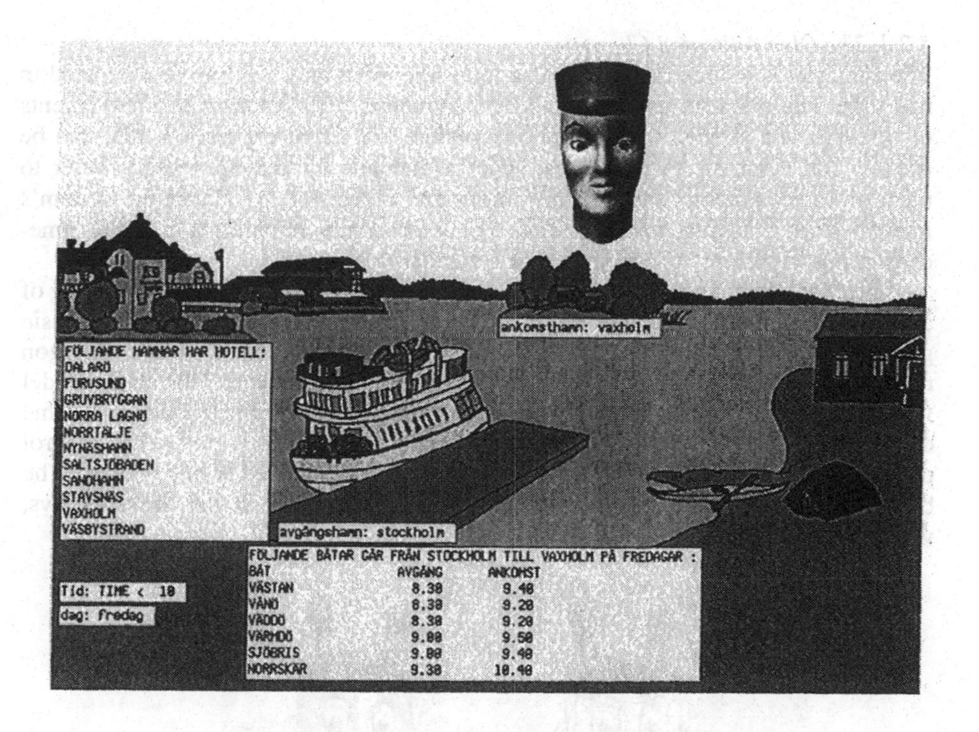

Figure14. The graphical model of the WAXHOLM micro-world.

5.2. The Olga Project

In the Olga project, we developed a multi-modal system combining a dialogue interface with a graphical user interface, which provided users with consumer information about microwave ovens. Feedback from the system is expressed through an animated talking character (Beskow, Elenius & McGlashan, 1997; Beskow & McGlashan, 1997).

Compared with previous research in the area, the novelty of Olga was that it integrated interactive spoken dialogue, 3D animated facial expressions, full body gestures, lip-synchronised audio-visual speech synthesis and a graphical interface. The system itself is composed of four main components: a speech and language understanding component; a direct manipulation interface which provides graphical information and widgets for navigation; an animated talking character using text-to-speech synthesis; and a dialogue manager for co-ordinating interpretation and generation in these modalities. Here we focus our description on the animated character.

5.2.1. The Olga Animated Character

The character is a three-dimensional cartoon-like robot lady that can be animated in real time. The character uses text-to-speech synthesis with synchronised movements of lips, jaw and tongue. It also supports gesture and facial expression, that can be used to add emphasis to utterances, support dialogue turntaking, visually refer to other on-screen graphics such as illustrations and tables, and to indicate the system's internal state: listening, understanding, being uncertain, thinking (i.e. doing time-consuming operations such as searching a database), etc.

The Olga character (Figure 15) is implemented as a polygon model, consisting of about 2500 polygons that can be animated at 25 frames per second on a basic graphics workstation. The character was first created as a static polygon representation of the body and head, including teeth and tongue. This static model was then parameterised, using the general deformation parameterisation scheme described in Section 2.1. Not only articulatory parameters, but also control parameters for eyebrows, eyelids and smiling were implemented in this manner. The body was parameterised by introduction of rotational joints at the neck, elbows, wrists, fingers, etc.

Figure 15. Wireframe and shaded representations of the Olga character.

5.2.2. Speech, Expression and Gesture

One important reason for using a full body animated agent in a spoken dialog system is the expectation that it will contribute to the efficiency and ease of the interaction. Speech intelligibility is expected to increase, provided that mouth movements are properly modelled and that users can transfer their speech reading faculties to the cartoon figure. In the series of experiments described in Section 3 above, we found that the Olga-character increased overall intelligibility in noise from 30% for synthetic voice only, to 47% for the synthetic voice - synthetic face combination.

As with the other talking faces, speech articulation of Olga is controlled from our rule-based text-to-speech system. Once the parameter trajectories are calculated, the animation is carried out in synchrony with playback of the speech waveform, which in turn is generated by the formant synthesiser controlled from the same rule-synthesis framework, but set to speaker specific parameter values typical for the "cartoon voice".

Speech movements are calculated on an utterance-by-utterance basis and played back with high control over synchronisation. Body movements and non-speech facial expressions, on the other hand, place different requirements on the animation system. Say for example that we want the agent to dynamically change its expression during a user utterance, depending on the progress of the speech recognition. In this case, obviously utterance-by-utterance control will not do. The basic mechanism for handling these kinds of movements in the Olga system is the possibility to, at any specific moment, specify a parameter trajectory as a list of time-value pairs to be evaluated immediately. Using such trajectory commands, gesture templates can be defined by grouping several commands together as procedures in a high-level scripting language (Tcl/Tk). This allows for complex gestures, such as "shake head and shrug" or "point at graphics display", which require many parameters to be updated in parallel, to be triggered by a simple procedure call.

Since a general scripting language is used, gesture templates can also be parameterised by supplying arguments to the procedure. For example, a pointing gesture might take optional arguments defining direction of pointing, duration of the movement, degree of effort, etc. As another example, a template defining a "shrug" gesture can have a parameter for selecting one from a set of alternative realisations, ranging from a simple eyebrow movement to a complex gesture involving arms, head, eyebrows and mouth. During the course of the dialogue, appropriate gestures are invoked in accordance with messages sent from the dialogue manager. There is also an "idle loop", invoking various gestures when nothing else is happening in the system. The scripting approach makes it easy to experiment with new gestures and control schemes. This sort of template-based handling of facial expressions and gestures has proven to be a simple, yet quite powerful way of managing non-speech movements in the Olga system.

5.3. The August System

There is an obvious need for more naturalistic talking faces. Rather than moving to concatenative, video-realistic solutions we have elaborated the original parametric model as was explained in section 2.7. The Swedish author, August Strindberg, provided inspiration to create the animated talking agent used in a dialogue system that was on display during 1998 as part of the activities celebrating Stockholm as the Cultural Capital of Europe. This dialogue system made it possible to combine several domains, thanks to the modular functionality of the architecture. Each domain has its own dialogue manager, and an example based topic spotter is used to relay the user utterances to the appropriate dialog manager. In this system, the animated agent "August" presents different tasks such as taking the visitors on a trip through the Department of Speech, Music and Hearing, and giving street directions and also presenting short excerpts from the works of August Strindberg, when waiting for someone to talk to.

August was placed, unattended in a public area of Kulturhuset in the centre of Stockholm. One challenge is this very open situation with no explicit instructions being given to the visitor. A simple visual "visitor detector" makes August start talking about any of his knowledge domains.

When August is spoken to he will respond only when possible, otherwise he will ignore the request or say something related to it. The complete August system was constructed out of modules from the speech toolkit developed at The Centre for Speech Technology (CTT) at KTH. The August system is a platform for presentation of the speech technology research done at CTT. During its lifetime, the system consequently does not have a fixed configuration with certain functionality, rather the system is continuously updated with new modules such as a more advanced dialogue handler to enable more advanced communication.

One of the goals of the system was to serve as a platform for experiments on human-computer communication; specifically how different aspects of the animated talking agent affect interaction. Therefore, when designing the agent, it was considered of paramount importance that August should not only be able to generate convincing lip-synchronized speech, but also exhibit a rich and believable non-verbal behaviour. To facilitate this we have developed a library of gestures based on the gestures developed in the Olga project. This library consists of communicative gestures of varying complexity and purpose, ranging from primitive punctuators such as blinks and nods to complex gestures tailored for particular sentences (Table 2). They are used to communicate such non-verbal information as emotion, attitude, turntaking, and to highlight prosodic information in the speech (i.e. stressed syllables and phrase boundaries as described in section 4). The parameters used to signal prosodic information in our model are primarily eyebrow and head motion but also gestures such as eye widening and narrowing are also available.

Table 2. Examples of typical functions for different motions in facial gestures.

Motion involved in gesture	*Typical function*
Eye blinks	Word boundaries, emphasis, idle random blinking.
Eye rotation	Thinking, searching, pointing
Head nodding	Emphasis, turntaking
Head turning	Spatial references (e.g. "the bathroom is over there"), attitude
Eyebrow raising	Mark words in focal position, stressed syllables, questions, emotions
Eyebrow frowning	Thinking, disagreeing
Emotional expressions	Semantically motivated (See Figure 16 for examples)

5.3.1. Prosodic Gestures

Having the agent augment the auditory speech with non-articulatory movements to enhance accentuation has been found to be very important in terms of the perceived responsiveness and believability of the system. The guidelines for creating the prosodic gestures were to use a combination of head movements and eyebrow motion and maintain a high level of variation between different utterances. Experiments such as the one described in Section 4 above indicate that associating eyebrow movements with words in focal position can be a possible way of controlling prosodic facial motion. Potential problems with such automatic methods are that the agent may look nervous or unnaturally intense and that the eyebrow motion could become too predictable.

To avoid predictability and to obtain a more natural flow, we have tried to create subtle and varying cues employing a combination of head and eyebrow motion. A typical utterance from August can consist of either a raising of the eyebrow early in the sentence followed by a small vertical nod on a focal word or stressed syllable, or a small initial raising of the head followed by eyebrow motion on selected stressed syllables. A small tilting of the head forward or backward often highlights the ending of a phrase. A number of standard gestures with typically one or two eyebrow raises and some head motion were defined. The standard gestures work well with short responses such as, "Yes, I believe so," or "Stockholm is more than 700 years old."

5.3.2. Gestures of Emotion

To enable display of the agent's different moods, six basic emotions similar to the six universal emotions defined by Ekman (1979) were implemented (Figure 16), in a way similar to that described by Pelachaud, Badler & Steedman (1996). Due to the limited resolution in our current 3D-model, some of the face properties, such as *'wrinkling of the nose'* are not possible, and were therefore left out. Appropriate emotional cues were assigned to a number of utterances in the system, often paired with other gestures.

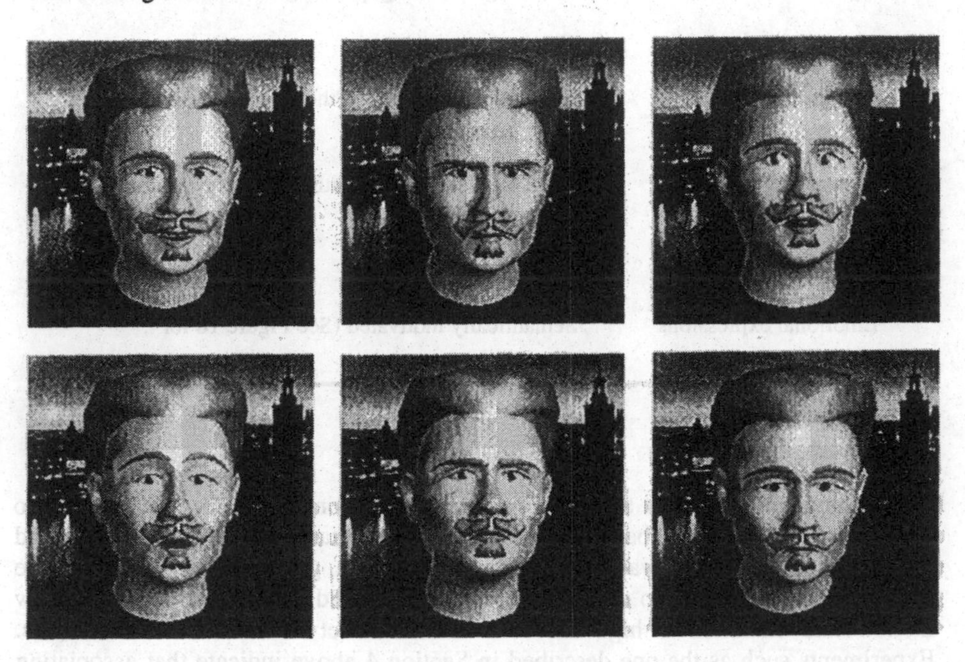

Figure 16. August showing different emotions (from top left to bottom right): Happiness, Anger, Fear, Surprise, Disgust and Sadness

5.3.3. An Example Sentence

Specially tailored, utterance specific gestures were created for about 200 sentences in the August system. An example of how eyebrow motion has been modelled in such a sentence is shown in Figure 17. The utterance is a Strindberg quotation "Regelbundna konstverk bli lätt tråkiga liksom regelbundna skönheter; fullkomliga människor eller felfria äro ofta odrägliga." (Translation: *"Symmetrical works of art easily become dull just like symmetrical beauties; impeccable or flawless people are often unbearable."*). Eyebrows are raised on the syllables *'-verk'*, *'-het'*, *'-er'*, *'-skor'* *'-fria'* and there is a final long rise that peaks on *'a'* in the last word *'odrägliga'*. Notice the lowering of the eyebrows that starts at t=5.8, which is not intended to convey prosodic information but rather emotional information. Not shown is the rotation of the head in the same utterance. At the first phrase boundary (after *'tråkiga'* at t=3.1 s), August tilts his head forward. At the next phrase boundary (before *'fullkomliga'* at t=5.5 s) he tilts the head even more forward and slightly sideways, lowers his eyebrows and looks very serious. A slow continuous raising of the head follows to the end of the utterance.

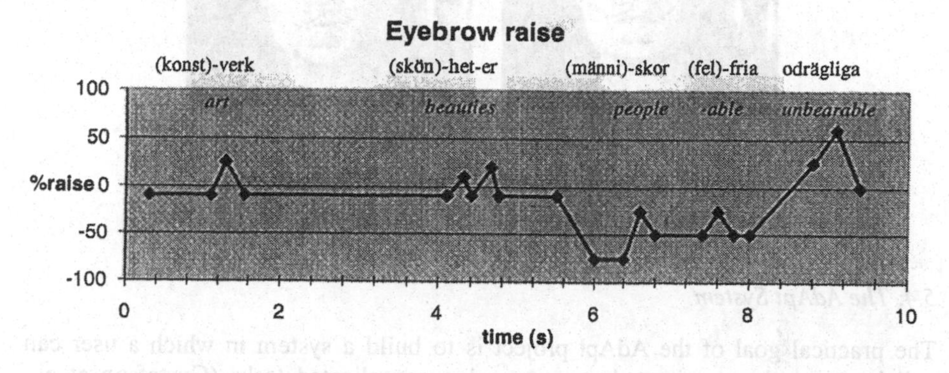

Figure 17. Example of eyebrow motion modelling in the August system. August says (in Swedish) "Regelbundna konstverk bli lätt tråkiga liksom regelbundna skönheter; fullkomliga människor eller felfria äro ofta odrägliga." (From Lundeberg & Beskow, 1999)

5.3.4. Listening, Thinking and Turntaking

For turntaking issues, visual cues such as raising of the eyebrows and tilting of the head slightly at the end of question phrases were created. Visual cues were also used to further emphasize the message (e.g. showing directions by turning the head). To enhance the perceived responsiveness of the system, a set of listening gestures and thinking gestures was created. When the user presses the push-to-talk button, the agent immediately starts a randomly selected listening gesture, for example raising the eyebrows. At the release of the push-to-talk button, the agent changes to a randomly selected thinking gesture such as frowning or looking upwards with the eyes searching. (Figure 18)

To catch the attention of the visitors to the exhibition, various entertaining behaviours of the agent were needed. For example, August can roll his head in various directions, he can pretend he is watching a game of tennis, and he can make a flirting gesture by looking toward the entrance to the exhibition and whistling while raising his eyebrows. These idling motions are important in making the system look alive, as opposed to a stationary, staring face that may make the system look "crashed" or "hung". A high degree of variation for these idling motions is needed to prevent users from predicting the actions of the dialogue system.

Figure 18. August listening (left) and thinking (right).

5.4. The AdApt System

The practical goal of the AdApt project is to build a system in which a user can collaborate with an animated agent to solve complicated tasks (Gustafson et al., 2000). We have chosen a domain in which multimodal interaction is highly useful, and which is known to engage a wide variety of people in our surroundings, namely, finding available apartments in Stockholm. In the AdApt project, the agent has been given the role of asking questions and providing guidance by retrieving detailed authentic information about apartments. The user interface can be seen in Figure 19.

Because of the conversational nature of the AdApt domain, the demand is great for appropriate interactive signals (both verbal and visual) for encouragement, affirmation, confirmation and turntaking (Cassell et al., 2000b; Pelachaud, Badler & Steedman, 1996). As generation of prosodically grammatical utterances (e.g. correct focus assignment with regard to the information structure and dialogue state) is also one of the goals of the system it is important to maintain modality consistency by simultaneous use of both visual and verbal prosodic and conversational cues (Nass & Gong, 1999). As described in Section 2.6, we are at present developing an XML-based representation of such cues that facilitates description of both verbal and visual cues at the level of speech generation. These cues can be of varying range covering attitudinal settings appropriate for an entire sentence or conversational turn or be of a shorter nature like a qualifying comment to something just said. Cues relating to turntaking or feedback need not be associated with speech acts but can occur during breaks in the conversation. Also in this case, it is important that there is a one-to-many relation between the symbols and the actual gesture implementation to avoid stereotypic agent behaviour. Currently a weighted random selection between different realizations is used.

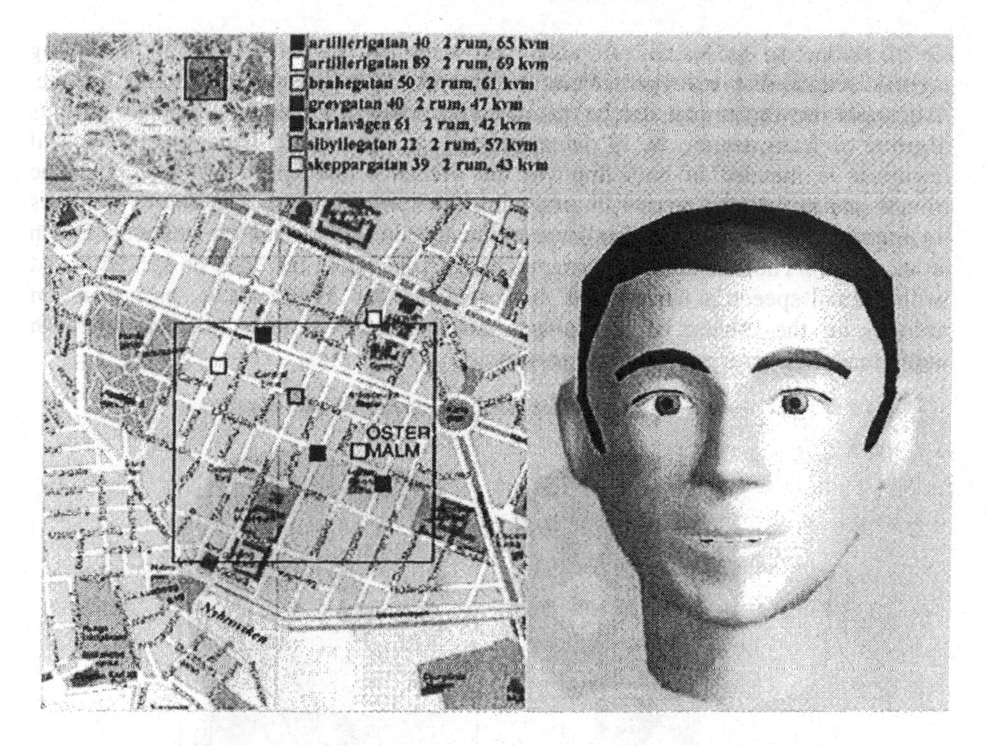

Figure 19. The agent Urban in the AdApt apartment domain.

5.5. An Automatic Language Tutor

The use of animated talking agents as automatic language tutors is an interesting future application that puts heavy demands on the interactive behaviour of the agent (Beskow et al., 2000). In this context, conversational signals do not only facilitate the flow of the conversation but can also make the actual learning experience more efficient and enjoyable.

The effectiveness of language teaching is often contingent upon the ability of the teacher to create and maintain the interest and enthusiasm of the student. The success of second language learning is also dependent on the student having ample opportunity to work on oral proficiency training with a tutor. The implementation of animated agents as tutors in a multimodal spoken dialogue system for language training holds much promise towards fulfilling these goals. Different agents can be given different personalities and different roles, which should increase the interest of the students. Many students may also be less bashful about interacting with an agent who corrects their pronunciation errors than they would be making the same errors and interacting with a human teacher. Instructions to improve pronunciation often require reference to phonetics and articulation in such a way that is intuitively easy for the student to understand. An agent can demonstrate articulations by providing sagittal sections that reveal articulator movements normally hidden from the outside. Articulator movement can also be displayed by rendering model surfaces in various degrees of transparency as is demonstrated in Figure 20. This type of visual feedback is intended to both improve the learner's perception of new language sounds and to help the learner in producing the corresponding articulatory gestures by internalising the relationships between the speech sounds and the gestures (Badin et al., 1998). The articulator movements of such an agent can also be synchronised with natural speech at normal and slow speech rates. Furthermore, pronunciation training in the context of a dialogue automatically includes training of both individual phonemes and sentence prosody.

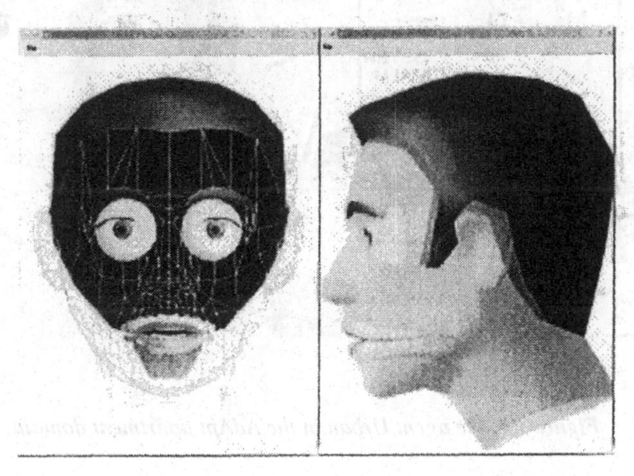

Figure 20. Examples of different display possibilities for the talking head model.

In learning a foreign language, visual signals may in many contexts be more important than verbal signals. During the process of acquiring a language, both child L1 speakers and adult L2 speakers rely on gestures to supplement their own speech production (McNeill, 1992; Gullberg, 1998). Adult L2 speakers often make more extensive use of gestures than L1 speakers, especially when searching for words or phrases in the new language. In this context, gestures have a compensatory function in production, often substituting for an unknown word or phrase. L2 listeners may also make greater use of visual cues to aid the conversational flow than do L1 listeners.

It has been found that the integration of segmental audio-visual information is affected by the relationship between the language of the speaker and that of the listener. Subjects listening to a foreign language often incorporate visual information to a greater extent than do subjects listening to their own language (Kuhl et al., 1994; Burnham & Lau, 1999). Furthermore, in a conversation, the L2 learner must not only concentrate on segmental phonological features of the target language while remembering newly learned lexical items, but must also respond to questions at the same time. This task creates a cognitive load for the L2 listener that is in many respects much different from that for the L1 user of a spoken dialogue system. Thus, the compensatory possibilities of modality transforms and enhancements of the visual modality are well worth exploring not only concerning segmental, phoneme-level information but also for prosodic and conversational information.

6. CONCLUSIONS AND PERSPECTIVES

This chapter has presented a method of audio-visual speech synthesis based on parametrically controlled 3D polygon models. This method has provided us with a powerful and flexible tool for audio-visual speech research as well as comprising a functional talking head for use in various experimental spoken dialog systems. There is currently much interest in exploring more sophisticated models and theories of speech production to increase the precision of synthesized visual articulatory movements. Interest in talking-head animation using the MPEG-4 standard for multimedia applications developed by MPEG (Moving Picture Experts Group) (Sikora, 1997) is also on the increase, which demonstrates the perceived need for standards in this area. New developments in animation techniques can also lead to increased naturalness for talking heads. Further research is needed to increase our knowledge on topics such as the time synchronization of facial gestures for prosody and communicative signals, and also how such gestures are integrated perceptually with the audio signal. In dialog systems, increased knowledge is needed concerning the coordination of gestures for turntaking and feedback including gestures for dealing with errors occurring during the dialog. We believe that the implementation of such knowledge will continue to improve the naturalness and utility of talking faces for spoken dialog systems in a multitude of future domains and applications.

7. REFERENCES

Agelfors, E., J. Beskow, M. Dahlquist, B. Granström, M. Lundeberg, K.-E. Spens & T. Öhman "Synthetic faces as a lipreading support". In: *Proceedings of ICSLP'98*, Sydney, Australia, 1998.

Agelfors E, J. Beskow, M. Dahlquist, B. Granström, M. Lundeberg, G. Salvi, K.-E. Spens & T. Öhman "Synthetic visual speech driven from auditory speech", In: *Proceedings of AVSP '99*, 123-127, Santa Cruz, USA, 1999.

Badin P, G. Bailly & L.-J. Boë. "Towards the use of a virtual talking head and of speech mapping tools for pronunciation training". In: *Proceedings of ESCA Workshop on Speech Technology in Language Learning (STiLL 98)*, 167-170, Stockholm: KTH, 1998.

Bertenstam J., J. Beskow, M. Blomberg, R. Carlson, K. Elenius, B. Granström, J. Gustafson, S. Hunnicutt, J. Högberg, R. Lindell, L. Neovius, A. de Serpa-Leitao, L. Nord & N. Ström. "The Waxholm system - a progress report", In: *Proceedings of Spoken Dialogue Systems*, Vigsø, Denmark, 1995.

Beskow, J. "Animation of Talking Agents", In: Proceedings of AVSP'97, ESCA Workshop on Audio-Visual Speech Processing, Rhodes, Greece, 1997.

Beskow J. & S. McGlashan. "OLGA - A conversational agent with gestures". In: André E,(Ed.), *Proc of the IJCAI -97 Workshop on Animated Interface Agents: Making them Intelligent*, Nagoya, Japan, 1997.

Beskow J. & K. Sjölander. "WaveSurfer - a public domain speech tool", In: *Proceedings of ICSLP 2000*, Beijing, 2000.

Beskow J., Dahlquist M., Granström B., Lundeberg M., Spens K.-E. & Öhman T. "The teleface project - multimodal speech communication for the hearing impaired". In: Proceedings of Eurospeech '97, Rhodos, Greece, 1997.

Beskow J., K. Elenius & S. McGlashan. "Olga - A dialogue system with an animated talking agent", Proceedings of EUROSPEECH'97,Rhodes,Greece. 1997.

Beskow J., B. Granström, D. House & M. Lundeberg. "Experiments with verbal and visual conversational signals for an automatic language tutor". In: *Proceedings of InSTiL 2000*, Dundee, Scotland, 2000.

Bregler C., M. Covell & M. Slaney. "Video Rewrite: Visual speech synthesis from video". In: *Proceedings of the ESCA Workshop on Audiovisual Speech Processing*, Rhodes, Greece, 1997.

Brooke N.M. & S.D. Scot. "An audio-visual speech synthesiser". *Proceedings of STiLL, The ESCA Workshop on Speech Technology in Language Learning*, Marholmen, Sweden, 1998.

Bruce, G., B. Granström & D. House. "Prosodic phrasing in Swedish speech synthsis". In: Bailly, G., C. Benoit & T.R. Sawallis (Eds.), *Talking Machines: Theories, Models, and Designs*, 113-125. Amsterdam: North Holland, 1992.

Burnham D. & S. Lau. "The integration of auditory and visual speech information with foreign speakers: The role of expectancy" In: *Proceedings of AVSP '99*, 80-85, Santa Cruz, USA, 1999.

Carlson R., B. Granström & S. Hunnicutt. "Multilingual text-to-speech development and applications". A.W. Ainsworth (Ed.), *Advances in speech, hearing and language processing*, London: JAI Press, UK, 1991.

Cassel, J., M. Steedman, N. Badler, C. Pelachaud, M. Stone, B. Douville, S. Prevost & B. Achorn. "Modeling the Interaction between Speech and Gesture", In: *Proceedings of 16th Annual Conference of the Cognitive Science Society*, Georgia Institute of Technology, Atlanta, USA, 1994.

Cassell J., J. Sullivan, S. Prevost & E. Churchill (Eds). *Embodied Conversational Agents*, Cambridge MA: The MIT Press, 2000a.

Cassell J., T. Bickmore, L. Campbell, V. Hannes & H. Yan. "Conversation as a System Framework: Designing Embodied Conversational Agents". In: Cassell J., J. Sullivan, S. Prevost & E. Churchill (Eds). *Embodied Conversational Agents*, Cambridge MA: The MIT Press, 2000b.

Cavé, C., I. Guaïtella, R. Bertrand, S. Santi, F. Harlay & R. Espesser "About the relationship between eyebrow movements and F0 variations". In: Bunnell, H.T. & W. Idsardi. (Eds). *Proceedings ICSLP 96*, 2175-2178, Philadelphia, PA, USA, 1996.

Cohen, M.M. & D.W. Massaro. "Modeling coarticulation in synthetic visual speech". In: N.M. Thalmann & D. Thalmann (Eds.) *Models and Techniques in Computer Animation*. Tokyo: Springer-Verlag, 139-156, 1993.

Ekman P. "About brows: Emotional and conversational signals". In: von Cranach M., K Foppa, W. Lepinies & D. Ploog. (Eds). *Human ethology: Claims and limits of a new discipline: Contributions to the Colloquium*, 169-248, Cambridge: Cambridge University Press, 1979.

Engwall O. "Making the Tongue Model Talk: Merging MRI & EMA Measurements", In: *Proc of Eurospeech 2001*, 261-264, 2001.

Ezzat T. & P. Tomaso. "MikeTalk: A talking facial display based on morphing visemes", *Proceedings of the Computer Animation Conference*, Philadelphia, PA, 1998.

Fant, G. & A. Kruckenberg,. ""Preliminaries to the study of Swedish prose reading and reading style. *STL-QPSR,KTH*, 2/1989, 1-80, 1989.

Gullberg M. *Gesture as a communication strategy in second language discourse. A study of learners of French and Swedish*, Lund: Lund University Press, 1998.

Gustafson, J., L. Bell, J. Beskow, J. Boye, R. Carlson, J. Edlund, B. Granström, D. House & M. Wirén "AdApt - a multimodal conversational dialogue system in an apartment domain". In: *Proceedings of ICSLP 2000*, (2) 134-137. Beijing, China, 2000.

Katashi, N. & T. Akikazu. "Speech Dialogue with Facial Displays: Multimodal Human-Computer Conversation", *Proceedings of the 32nd Annual Meeting of the Association for Computational Linguistics (ACL-94)*, pp. 102-109, 1994.

Kuhl P.K., M. Tsuzaki, Y. Tohkura & A.M. "Meltzoff Human processing of auditory-visual information in speech perception: Potential for multimodal human-machine interfaces", In: *Proceedings ICSLP '94*, 539-542, Yokohama, Japan, 1994.

Lundeberg, M. "Multimodal talkommunikation - Utveckling av testmiljö", Master of science thesis (in Swedish). TMH-KTH, Stockholm, Sweden, 1997.

Lundeberg, M. & J. Beskow. "Developing a 3D-agent for the August dialogue system", In: *Proceedings of AVSP'99*, 151-156, Santa Cruz, USA, 1999.

MacLeod A. & Q. Summerfield. "A procedure for measuring auditory and audio-visual speech reception thresholds for sentences in noise. Rationale, evaluation and recommendations for use". *British Journal of Audiology*, 24:29-43, 1990.

Massaro, D.W. *Perceiving Talking Faces: From Speech Perception to a Behavioural Principle*. Cambridge, MA: MIT Press, 1998.

McNeill D. *Hand and mind: What gestures reveal about thought*, Chicago: University of Chicago Press, 1992.

Nass C. & L. Gong. "Maximized modality or constrained consistency?" In: *Proceedings of AVSP'99*, 1-5, Santa Cruz, USA, 1999.

Neely, K.K. "Effects of visual factors on intelligibility of speech". *Journal of the Acoustical Society of America*, 28, 1276-1277, 1956.

Parke F.I. "Parametrized models for facial animation". *IEEE Computer Graphics*, 2(9), pp 61-68, 1982.

Pelachaud C., N.I. Badler & M. Steedman. "Generating Facial Expressions for Speech". *Cognitive Science 28*, 1-46, 1996.

Pelachaud, C., E. Magno-Caldognetto, C. Zmarich & Cosi P. "An approach to an Italian talking head". In: *Proceedings of Eurospeech 2001*, Aalborg, Denmark, 2001.

Platt, S.M. & N.I. Badler. "Animating Facial Expressions". *Computer Graphics*, Vol. 15, No. 3, pp. 245-252, 1981.

Sikora, T. "The MPEG-4 Video Standard and its Potential for Future Multimedia Applications". In: *Proceedings IEEE ISCAS Conference*, Hongkong, 1997.

Terzopoulos, D. & K. Waters. "Physically based facial modelling, analysis and animation". *Visualisation and Computer Animation*, 1:73-80, 1990.

Thórisson, K.R. "A Mind Model for Multimodal Communicative Creatures & Humanoids". *International Journal of Applied Artificial Intelligence*, Vol. 13 (4-5), 449-486, 1999.

Waters, K. "A muscle model for animating three-dimensional facial expressions", *Computer Graphics*, 21:17-24, 1987.

Öhman T. "An audio-visual database in Swedish for bimodal speech processing". *TMH-QPSR, KTH*, 1/1998.

8. AFFILIATIONS

Björn Granström, David House & Inger Karlsson
Department of Speech, Music and Hearing
KTH
Drottning Kristinas Väg 31,
SE-104 01 Stockholm, Sweden
http://www.speech.kth.se

Text, Speech and Language Technology

1. H. Bunt and M. Tomita (eds.): *Recent Advances in Parsing Technology.* 1996 ISBN 0-7923-4152-X
2. S. Young and G. Bloothooft (eds.): *Corpus-Based Methods in Language and Speech Processing.* 1997 ISBN 0-7923-4463-4
3. T. Dutoit: *An Introduction to Text-to-Speech Synthesis.* 1997 ISBN 0-7923-4498-7
4. L. Lebart, A. Salem and L. Berry: *Exploring Textual Data.* 1998 ISBN 0-7923-4840-0
5. J. Carson-Berndsen, *Time Map Phonology.* 1998 ISBN 0-7923-4883-4
6. P. Saint-Dizier (ed.): *Predicative Forms in Natural Language and in Lexical Knowledge Bases.* 1999 ISBN 0-7923-5499-0
7. T. Strzalkowski (ed.): *Natural Language Information Retrieval.* 1999 ISBN 0-7923-5685-3
8. J. Harrington and S. Cassiday: *Techniques in Speech Acoustics.* 1999 ISBN 0-7923-5731-0
9. H. van Halteren (ed.): *Syntactic Wordclass Tagging.* 1999 ISBN 0-7923-5896-1
10. E. Viegas (ed.): *Breadth and Depth of Semantic Lexicons.* 1999 ISBN 0-7923-6039-7
11. S. Armstrong, K. Church, P. Isabelle, S. Nanzi, E. Tzoukermann and D. Yarowsky (eds.): *Natural Language Processing Using Very Large Corpora.* 1999 ISBN 0-7923-6055-9
12. F. Van Eynde and D. Gibbon (eds.): *Lexicon Development for Speech and Language Processing.* 2000 ISBN 0-7923-6368-X; Pb: 07923-6369-8
13. J. Véronis (ed.): *Parallel Text Processing.* Alignment and Use of Translation Corpora. 2000 ISBN 0-7923-6546-1
14. M. Horne (ed.): *Prosody: Theory and Experiment.* Studies Presented to Gösta Bruce. 2000 ISBN 0-7923-6579-8
15. A. Botinis (ed.): *Intonation.* Analysis, Modelling and Technology. 2000 ISBN 0-7923-6605-0
16. H. Bunt and A. Nijholt (eds.): *Advances in Probabilistic and Other Parsing Technologies.* 2000 ISBN 0-7923-6616-6
17. J.-C. Junqua and G. van Noord (eds.): *Robustness in Languages and Speech Technology.* 2001 ISBN 0-7923-6790-1
18. R.H. Baayen: *Word Frequency Distributions.* 2001 ISBN 0-7923-7017-1
19. B. Granström, D. House and. I. Karlsson (eds.): *Multimodality in Language and Speech Systems.* 2002 ISBN 1-4020-0635-7

KLUWER ACADEMIC PUBLISHERS – DORDRECHT / BOSTON / LONDON